Without Hierarchy

Without Hierarchy

The Scale Freedom of the Universe

MARIAM THALOS

OXFORD
UNIVERSITY PRESS

OXFORD
UNIVERSITY PRESS

Oxford University Press is a department of the University of Oxford.
It furthers the University's objective of excellence in research,
scholarship, and education by publishing worldwide.

Oxford New York
Auckland Cape Town Dar es Salaam Hong Kong Karachi
Kuala Lumpur Madrid Melbourne Mexico City Nairobi
New Delhi Shanghai Taipei Toronto

With offices in
Argentina Austria Brazil Chile Czech Republic France Greece
Guatemala Hungary Italy Japan Poland Portugal Singapore
South Korea Switzerland Thailand Turkey Ukraine Vietnam

Oxford is a registered trade mark of Oxford University Press
in the UK and certain other countries.

Published in the United States of America by
Oxford University Press
198 Madison Avenue, New York, NY 10016

Library of Congress Cataloging-in-Publication Data
Thalos, Mariam, 1962–
Without hierarchy : the scale freedom of the universe / Mariam Thalos.
pages cm
Includes bibliographical references.
ISBN 978-0-19-991764-8
1. Causation. 2. Science—Philosophy. 3. Physics—Philosophy. I. Title.
Q175.32.C38T43 2013
501—dc23 2012045539

For my son Oliver Ethan Richardson, who has the heart of a mathematician and a soul that hungers to glimpse all the intelligible world.

CONTENTS

ACKNOWLEDGMENTS

This book has benefitted from the wisdom of many people over the several decades during which its component parts have been a work in progress. My feeble gesture here at formal thanks to individuals by name will doubtless miss at least one someone who has played an important part in fostering or developing the ideas in this book. And for that I am deeply sorry. The most subtle influences are often the hardest to remember, despite being the most pervasive, while bold cautionary remarks and broad encouragement bound easily to mind. And sometimes the thanks one is due comes too late; two of the people I shall name here can no longer be thanked properly. Still, it does good to give thanks where thanks are due, and it will be my pleasure to do that by beginning with the present and working backwards.

I want to thank colleagues at the University of Utah who have indulged my periodic and quite often incoherent ramblings on the subjects of this book: Chrisoula Andreou, Stephen Downes, Leslie Francis, Matt Haber, Bruce Landesman, Lije Millgram, Clif McIntosh, Anya Plutynski, Jonah Shupbach, Cindy Stark and Nick White; and thanks too to Ron Mallon, Ram Neta, and Shaun Nichols, now former colleagues. I owe a special debt to Barry Smith, my first colleague-mentor at the University at Buffalo. His minimalist but carefully calculated style of guidance has been indispensable in my development as a writer and professional philosopher. Newton Garver, Jorge Gracia, Peter Hare, and John Kearns also gave freely of friendship and support to their junior colleague in Buffalo.

My first teachers in philosophy (to which discipline I arrived late in my student career) played an early role in the development of my ideas about science and its dignity. Thanks to Rolf Eberle, Henry Kyburg, Jr., and Deborah Modrak, without whose kindness as people and teachers I should never have learned to appreciate the histories shaping fundamental conceptions. And a special thanks to Mark Wilson who taught me that one sees something new every time one revisits a text, even one's own.

Colleagues in the larger philosophical community have also played important parts in encouraging, guiding, and shaping the ideas in this book. I would like to thank Craig Callendar, Nancy Cartwright, Bob Goodin, Kathrin Koslicki, Don Ross, Jonathan Schaffer, Ted Sider, Jessica Wilson, and John Woods for their serious and thoughtful engagement with those ideas of my book that they encountered piecemeal. Several anonymous referees for Oxford provided sage advice at crucial points in the process. For providing inspiration, I would like to thank Steven Strogatz and Lazlo Barabasi for their masterful teaching across disciplinary boundaries in books for wider audiences.

None of these kind people should be held to account for the present condition of the book or the ideas in it, but there is no question that the book is in a better place now because each of them played some part in my life.

The Institute of Philosophy at the University of London provided me a precious interval in the summer of 2010, away from pressures of ordinary life, in which to conceive the very idea of putting the pieces here assembled together in one place. This book has been a kind of growing up for the numerous ideas that have populated my thinking; many of them found their way into this book as it matured. Many of the ideas congregated here appeared first somewhere else:

"Degrees of Freedom: An Essay on Competitions between Micro and Macro in Mechanics," *Philosophy and Phenomenological Research*, **59**:1–39 (1999).

"Explanation is a Genus: On the Varieties of Scientific Explanation," *Synthese*, **130**:317–354 (2002).

"Nonreductive Physics," *Synthese*, ***149***:133–178 (2006).

"Systems," *Monist*, **92**(3):452–478 (2009).

"Two Conception of Fundamentality," *Philosophy of the Social Sciences*, 41(2):151–177 (2011).

"In Favor of Being Only Humean," *Philosophical Studies*, **93**:265–298 (1999).

"Molecule-for-Molecule Duplication," *Teorema*, 28(3):103–114 (2008).

"The Reduction of Causation," in M. Thalos and H. Kyburg, Jr., eds., *Probability is the Very Guide of Life: The Philosophical Uses of Probability*, Open Court, (2003).

"Truth Deserves to be Believed," *Philosophy*, 88:344 (2013).

Writing this book has also been a growing up experience for me as a human being. I could not have remained the course without the support of my husband

Robert David Richardson and my sons, Oliver Richardson and Elijah Thalos. Just their nearness buoyed me up throughout.

My research assistants at the University of Utah deserve a special word of thanks for their dedication and toil under time and other pressures: thank you, Kimberly Johnston, Anna Vaughn and Edward Andrew Greetis.

Peter Ohlin, philosophy editor at Oxford, deserves a special word of gratitude. Thank you, Peter, for seeing the gleam in the rough, and allowing me to work to bring out a larger shine.

Without Hierarchy

Introduction

Toward a Theory of Science

> [T]here is something arrogant and warlike in the notion of a man setting the universe in order. It is a kind of balance that is his glory. He rides the drifts like an escaped ski. His course is the caress of the hill. Far from flying with the angels, he traces with the fidelity of a seismograph needle the state of the solid bloody landscape.
>
> —Leonard Cohen, *Beautiful Losers* (1966)

1. Concerning Introductions

The first complete draft of this book began with materials that now lie in the hinterlands of its chapter 7. But there is nothing special about this book in that regard. As a book's readership grows and diversifies, and as the author comes to *anticipate* further widening, the author (in the hands of a good editor) comes to appreciate that the work's "core" might require further and further contextualization. In this way also older works get recontextualized for more contemporary readerships. With each reissue of a classic work, a preface is written, or a new introduction by a contemporary figure, ensheathing the original work in ever more recontextualizations. If a new introduction fulfills its function, the audience taking in the book grows even wider, making it likely that still more recontextualizations will be produced for the next generations of readers. Often the new "skins" provide historical materials of interest to any reader, but which the reader of the original work would not have been fortunate enough to access by way of an introduction—for instance, information about the work's critical reception or its place in a given literature or canon. Just as often, these new layers of material quote the work itself, as if to foster an innocent reader's appreciation of the material by providing in advance much-needed interpretation or stage-setting.

Contextualizations of things are not inert vis-à-vis their originals. If they were, they'd serve no purpose whatsoever. The reclothing of the old sometimes makes of it something brand new. This idea shall be a recurring theme in the present book—a model for how new things appear in the universe. True, the universe might have started with the simplest and most elemental things. But now the world is full of much more intricate things. How does this happen? My answer, as it will unfold time and again in this book, will follow its own pattern of resheathing. The answer is that, not only do new things come into existence by way of the familiar process of cobbling together simpler things already in existence—elements or building blocks that are plentiful and largely identical to one another because of it—such as when an author drafts something new; but *also* that new things come into existence through the process of reclothing old things in new sheaths. The former is like telling a story for the first time; the latter is much more like a quotation, such as occurs when one devises a story, commentary, or theory *about* characters already in existence. We recognize the latter as belonging to a different category of creation altogether, and we hold it to different standards of assessment.

It is the ultimate destination and key concern of this book to articulate clearly and in as many ways as possible the difference between cobbling and sheathing and also to articulate how the old features of an object give way to new ones during the sheathing process while never actually disappearing. The clear articulation of this idea will require analyzing and divesting ourselves of certain entrenched metaphysics, and in the process introducing new metaphysics. (We will be resheathing our very understanding!) This process also requires that we work up to an explicit analysis of resheathing by way of a scenic route through a very wide range of topics, to supplement and enrich the philosopher's usually restricted diet of examples.

An important point I will be seeking to establish is that resheathing is a *very different* thing, speaking metaphysically, from cobbling. Cobbling is a simple putting together; where by contrast sheaths provide added structure, while the ensheathed continues to provide some but not all of what it provided before being resheathed. More importantly, sheathing and cobbling are things that occur at different scales of action. Obviously what I need to do to convince you of any of this is to explain what a sheath is. I can't do that without some philosophical preliminaries to clear away some misconceptions and false dogmas. So let my remarks thus far serve as promissory notes for what is in store; remarks on resheathing will proceed for some time as a work in progress. And I intend to use "sheathing" and "resheathing" interchangeably, to mean simply the provisioning of a new sheath for something that may or may not already possess a sheath.

What does the resheathing process look like, very roughly? We already know what this process looks like, at least from a distance, because at our scale of existence we employ it and interact with its products; they're all around us: flocking birds, human and nonhuman societies, and (as I'll explain in due course) magnets and superconducting materials. Indeed, we ourselves as living things are such products. Biologists have a term for resheathings that occur on their watch: they call them *recruitments*; cognitive scientists have also begun to use this "recruitment" language. Both groups of scientists argue that evolution and culture use materials already present in an organism (sometimes together with materials in its environment) and build upon them to enlarge or alter their functionality within the organism or context. I have something a bit more general in mind by resheathing, although recruitment will qualify as a form of resheathing. My term is intended as a label for a category of activity that happens at every scale of measurement, every scale of organization in the universe, when an entity or process at one scale manages to organize or reorganize entities and processes at a somewhat smaller scale.

Resheathing is an exercise of power analogous to the exercise of legitimate political authority enjoyed by a sovereign within its borders in times of peace. Resheathing, in other words, shares features with the exercise of legitimate authority or leadership-features distinctive from those that characterize the exercise of brute force. It is evocative of the Leonard Cohen passage that stands at the head of this introduction (though Cohen was talking about saints). An entity exercising the powers of legitimate leadership rides, like Cohen's ski, on top of an existing social reality (the institutions of a state) that it manages, with the most slender of motions, to tip in just this or that direction. The power so exercised is only rarely manifested in shows of force, or even the threat of force. Legitimate authority is quite routinely held through nonviolent means: the muscle of a legitimate state is rarely if ever flexed against its citizens, in spite of the fact that the typical legitimate state enjoys a complete monopoly on the legitimate use of force within its borders.

Resheathing is much more like an exercise of legitimate authority than it is like an application of brute force. The core message of this book is that activity, and especially resheathing activity, is happening at every scale of the universe. The resheathing of old things explains why the universe is—and this is now new philosophical terminology that will become clearer as we proceed—*scale free*. We arrive now at the point where we need to start attending to those philosophical preliminaries I spoke of, in preparation for the journey.

2. The Picture of Scale Freedom

The picture of the world I will advance is this one: there is all manner of activity in the universe, happening at every scale. Some might call this causation. I shall argue (in chapter 4) that another term should be used, because the term "causation" is already gainfully employed: it is used to signify something completely different. I shall explain why we can't use the term for both uses, for they are incompatible. So I've chosen the term "activity." Please grant me the usage of "activity" for now, until it becomes clear later why we need it.

So the contention is that there is activity *at every scale*. This contention is very much at odds with what a preponderance of professional metaphysicians nowadays believe. (Scientists as a larger class that includes biologists, geologists, sociologists, and ecologists are another matter, and I don't mean to be characterizing their opinions here in any way.) The point is this: the contention that there is activity at every scale is at odds with the contention that there is activity at only one scale—that for need of a simple handle we might refer to simply as the *micro* scale. According to proponents of a certain prevailing view, the *micro* is either (1) the scale at which the (uniquely) most minute parts of the universe interact with each other, and so also the scale at which operate the rules by which they are governed in those interactions;[1] or (2) even if there is no single (unique) set of most minute components of the universe, *micro* refers to the scale *above* which nothing "fundamental" can possibly reside. So the prevailing view—that I shall be rejecting—is the view that the universe is active only at scales smaller than those we can inspect with the naked eye, in spite of its vast size, estimated at nearly 100 billion light-years, or 10^{27} meters.

To say (as I shall be doing unrelentingly against a dogma that erupts in different and multiple guises) that the universe exhibits activity at every scale of measurement is to say that there is no special or privileged scale at which to view activity in the universe. In other words, that the universe has no "true" scale to which activity is confined. In other words still, the universe is *scale free*. I don't mean by that that the universe has no "bottom"—no minutest scale; it might well have. That's a different matter. I mean simply that when we are counting the number of scales at which there is activity in the universe, we cannot stop at one. My opponents are those who say that this is a mistake, that indeed there is a special and privileged scale (designated as the scale of micro physics—as if there were such a unique scale, which is doubtful, as will soon become clear). The critic will say that viewing the universe at larger scales obscures in important ways the fundamental activities of the universe, that the best we can hope for, by way of descriptions at higher scales, is a coarse-grained analysis or description of the true goings-on "down there." From the higher-scale vantage point, the critic continues, the true goings-on in the universe are "ciphered" and thus require decoding: this, in the critic's view, is the work of scientific analysis. The dispute between myself and my opponents is a dispute in metaphysics. It concerns the relationship between the micro and the macro. My opponents adhere to a form of *ontological reduction*. Thus I am opposing a version of reductionism. My own model of the universe is a version of antireductionism.

[1] Some would use the term "fundamental" here. I shall be advocating for another sense of "fundamental" in chapter 3.

I am using the term *reductionism* here very generously. By it I am referring to views that embrace the doctrine according to which there is a Master Science—which I'll sometimes refer to simply as "MS." This might not be Physics. (Of course everyone knows it *must* be Physics. And we mustn't pretend that alternatives to Physics have received genuine consideration since the late nineteenth century. Still, logically speaking, from the perspective of this metaphysical doctrine of reduction, it *needn't* be Physics.) From the reductionist's perspective, the universe has a single scale, though it may be viewed or grasped at other scales in "ciphered" form. The MS contains the key to all the ciphers. We require some translation from the MS to the ciphered presentations, some more complex than others, but rest assured, they exist. Because the MS, once completed, will contain the true Theory of Everything (TOE). If someone wishes to get to the "bottom" of some phenomenon, they are required to query the TOE—provided of course that the MS is not still a work in progress. The reductionist works with a hierarchical picture of the ontology of the universe. Figure 0.1 displays that picture.

This hierarchical picture comes from chapter 7—where I will criticize the doctrine associated with it in rather more mathematical language and present a more general argument than those you will be encountering in the early chapters of the book. I advise the reader regarding this "quotation" of the image (figure 0.1) to remind you of the resheathing process, which is incompatible with this picture of the ontology of the universe. This mainstream picture, embraced by reductionists *and also* by a vast preponderance of their contemporary critics alike, makes a statement to the effect that the only independently existing entities exist at the "ground floor," and everything else is composed of these as constituents, in ever more increasing levels of organization. As long as we stipulate, as in this picture, that the ontologies and processes of the "nonfundamental" levels are derivative, we can fulfill the promise of monism—only one kind of substance, and it is physical. No funny stuff—no ectoplasm, no disembodied

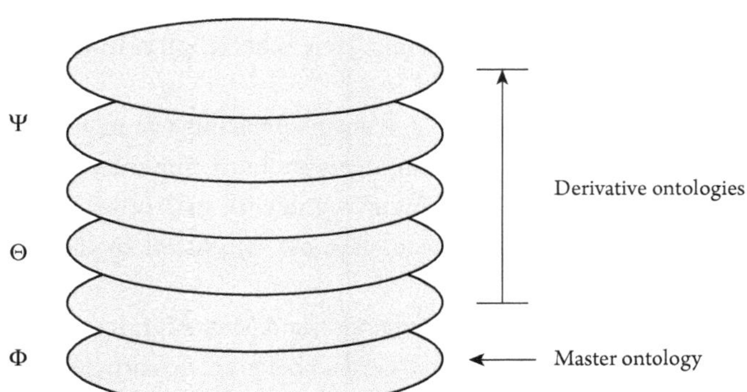

Figure 0.1 The contemporary reductionist ontology

minds, no voodoo, no monkey business of any kind. Just unsentimental, clear-eyed, no-nonsense science. Physicalists whose primary concern is for preserving monism simply insist on the priority of the ontology of Physics. But once the priority of the ontology of Physics has been stated, this immediately raises the question as to the independence of the other sciences, especially those most remote from Physics' paradigm case studies.

Once the conception of a Master Science is firmly anchored to the physicalist vision of the universe, there has sometimes been a clamor and a consternation regarding the so-called "autonomy of the sciences." Many scholars today undertake a reconciliation project—for the sake of rehabilitating something autonomy-like for so-called "special sciences" while nonetheless acknowledging the existence of an MS. There is now a large literature on intertheory relations that is meant to help us keep track of the relations between theories of the "special" sciences and those of the MS. (The MS is usually not called the "general" science, but that, of course, is the relevant contrast—after all, a TOE is plainly as general a science as one can have, if there's anything in a name.)

How are the "autonomous sciences" themselves depicted in this literature, in general terms? In almost every such account of the so-called *autonomy* enjoyed by these sciences vis-à-vis one another or the MS, they are almost always organized in a hierarchy, with the MS at a relevant extremum. In other words, sciences are organized by the size scale of their subject (of course with the assumption that there is such a unique scale), with smaller being closer to MS, hence more fundamental. The picture of the sciences, in these reconciliation projects, mirrors the hierarchical ontology of the universe depicted in figure 0.1, in reverse: the MS is at the pinnacle of the hierarchy, because its subject matter is at the bottom of the size scale.

One thing to note in passing is that the hierarchy is a very, *very* modern model of the relations of the sciences—modern in the sense of new since the atomism of John Dalton—and did not exist in the time of René Descartes. Even Isaac Newton provides an interesting point of contrast with this notion of an MS. True: Newton conceived of Physics as a master science, but on his conception of it, this meant that Physics just would explain everything. Here is his vision of how it would do so:

> All these things being consider'd, it seems probable to me, that God in the Beginning form'd, Matter in solid, massy, hard, impenetrable, moveable Particles, of such Sizes and Figures, and with such other Properties, and in such Proportion to Space, as most conduced to the End for which he form'd them; . . .
>
> And since Space is divisible *in infinitum*, and Matter is not necessarily in all places, it may be also allow'd that God is able to create Particles of Matter of several Sizes and Figures, and in several Proportions to Space, and perhaps of different Densities and Forces, and thereby to vary the Laws of

> Nature, and make Worlds of several sorts in several Parts of the Universe. At least, I see nothing of Contradiction in all this. (1718, 375–76, 378–80)

Newton's idea that a fundamental particle can be of any size is foreign to the conception of physicalism that we find today—the physicalism that harbors hierarchies.

I reject the hierarchy outright. It is an abomination. At its absolute best, it presents us with a simple-minded conflation of size with scale of aggregation—the latter of which might (though positively cannot always) be construed as a measure of or guide to complexity. But size and complexity are very different things—as Newton's vision clearly demonstrates. The conflation does nothing but harm.

While it is not my picture of the sciences, and while I will introduce my own picture shortly, it is worthwhile looking at another picture of the relations among sciences that the physicist Sunny Auyang paints in her books. I like her pictures, though they are not entirely interchangeable with mine. They have virtues not shared by the mainstream hierarchical conception. She says, "Science is not like a skyscraper in which the upper floors rest on the lower; it is closer to an airport in which the concourses cooperate as equals in an integral system" (1998, ix). While the spirit of the image is correct, it suggests an absence of overlap among the sciences—or at any rate an absence of overlap among their objects of study. But overlap is really fundamental to understanding the universe, and so consequently it is really fundamental to the understanding of the sciences themselves and their historical development, so it is something I want to capture. On one thing Auyang and I are very much in agreement: the so-called *levels* of the contemporary physicalist's picture have to go.

My picture shares some philosophical aspects also with that of Nancy Cartwright (1989 and since). She maintains a certain pluralism, to the effect that there are different kinds of systems, and hence different kinds of scientific modeling is appropriate to them. I agree wholeheartedly on this point. Where I disagree with Cartwright is on the metaphysics of this difference. She maintains that the difference is in conceptions of causality—where the term "causality" is used as a kind of lingua franca among the sciences. I will be arguing (in chapter 4) that causality is no lingua franca. It has other uses, some of which Cartwright too wants captured under the umbrella term. But as I shall explain, that umbrella is too small. We need a variety of umbrellas in the philosophical closet. That too will be a theme of this book.

3. What Pictures Do

Pictures are virulent; they are highly infectious. Even when inexplicit themselves, or when they are explicitly known to be too simple, they still serve to guide explicit thought. And because the hierarchy picture is very simple, it is that

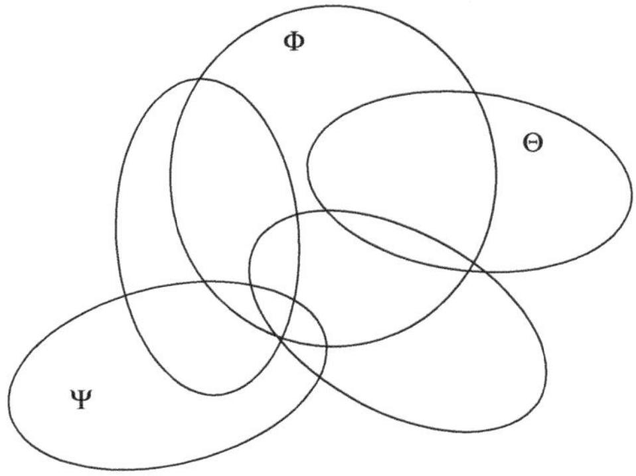

Figure 0.2 My picture, no hierarchies

much more virulent, that much harder to exorcise. To counteract a simple picture, one requires (ideally) an even simpler one. I offer you such a picture here (see figure 0.2). I can draw it without words, and nothing special needs to be said about any of the overlapping ovals. That, I fervently hope, will make the picture even simpler than the dominant picture of the sciences rendered in figure 0.1.

4. Some Other Pictures That Are Not Mine

The conceit of levels is attractive to many thinkers in many walks of the academy. There is a tidiness to it, with preassigned spaces for organizing things in what seems like a methodical, disciplined way: Physics, Chemistry, Biology, Neuroscience, Psychology, Ecology, Geology, Sociology, Cosmology. But in what order these last? How about Biochemistry and Neurobiology—do they have clear positions in this edifice? Where do the many forms of medicine and engineering belong? Why is Physics, as familiarly recognized in the academy, represented at both ends of the list? Never mind the specifics of the organization, obviously the thing to be admired in a hierarchical structure is its tidiness. (Unfortunately, the tidiness does not exist in practice. The messier picture I offer you is much more accurate.) Some philosophers, many of whom abhor reductionism almost as much as I do, wish to retain this hierarchy.

There are, after all, some compelling reasons for a picture with hierarchies. It has the quality of compositionality to it: first the basic building blocks of the universe come together to form a level-1 entity, then level-1 entities aggregate to form level-2 entities, and so on. And nothing new, in the way of activity, occurs at levels above the ground floor.

But this picture papers over the monumentally important fact that big things encounter little things all the time, one on one, as we might say, albeit "from above" and "from below" at the same time; and there are interesting interactions between the big and the little. Physics—*especially* Physics—discusses these, and quite frequently in terms that defy the language of minutest-on-minutest. We will of course be reminded by proponents of a one-scale universe that macro physics is just as much subject to being "disappeared," once the true scale of the universe comes into view, as any other discipline. Still, there is no proof of this principle, not even a shred of evidence—merely a kind of faith in it that borders on religion. It is essentially because there is no room in the hierarchy model for this monumentally important fact—that small interacts with big, well and truly—that this model is fatally flawed.

Those who find the compositionality of a hierarchy of levels pleasing might insist that the work done by levels is only to organize ontology, to aid in thinking about the constitution of objects, and to do this in large part mereologically—by examining the ways that parts contribute to the making of wholes. Many of these kinds of enthusiasts are not especially inclined toward the notion that the universe is one-scaled. They discuss "between-level" traffic, activity that transpires between different scales of objects. They suggest that the "upper levels" are constituted or diagnosable by this complexity of traffic in activity—not merely that there is this activity, but also that this complex traffic is what brings levels into being in the first place. I find this position quite mystifying, to say the least, and will discuss this point at some length in chapter 1. Suffice it to say here, by way of preview, that the complexity of traffic is in no way inconsistent with the one-scale vision of the universe that is the true target of my book. So it is unclear what "levels" are for. Complexities, on that understanding, all occur at the "true" scale of the universe. So it is in no way a problem with the one-scale model that complexity exists—if there is complexity, that is a further fact. The real problem with the doctrine that complexity creates levels, as I shall explain in chapter 1, is that it is metaphysically muddled. And so it is simply untenable.

5. What My Picture Combats

The one-scale universe is my target in this book. I have no intentions of defending dualisms of any kind—I believe that all that exists is physical, no matter what its scale. So organisms, societies (as such), and minds (as such) are all physical.*
Where I part company with contemporary physicalists is on this point: I simply don't think that all the action in the universe is restricted to micro-on-micro. I intend to present arguments as to why we should adopt the scale-free hypothesis. And the best means of doing away with dualisms (and it is indeed a *very*

* The concreta, as I shall explain.

simple philosophical proposal) is just to do away with level demarcations. The demarcations are meant to be dividing between kinds of entities too, as much as between regimes of order; some entities are marked for elimination or ontological demotion of some kind, while others are not so marked. Once there are no level demarcations, there are no means of distinguishing between licit and illicit kinds either—so there will be nothing marked as eligible for elimination or demotion (where ironically the "elevations" in the hierarchy correspond to metaphysical demotions). The level demarcations, as the rest of the book will explain, are metaphorical stumbling blocks to understanding the relations between sciences anyway, erecting barriers to interactions where there are none.

Before we embark, here is a very bare-bones argument against the position I shall occupy and defend—an argument that requires diffusing immediately. It is a distillation from arguments that I will deal with in more fine grain in chapter 1. From the point we dispose of this argument, we can work toward seeing why we require no replacement argument for the same conclusion. For (or so we will conclude by the end of the book) its conclusion is unwanted. So here now is that argument:

Let's assume that everything is organized into levels à la the picture in figure 0.1 Then begin with the commitment to monism:

1. Everything is physical.

By this we mean that everything can be characterized in physical terms, for example in terms of the parts and their physical interactions (right down to atoms, electrons, quarks, and what have you), even if it also has a manifestation that puts it in higher levels. Next:

2. Everything that has an effect on a physical thing must be having a physical effect on it at its minutest level.

That's part of what it means to have an effect on a physical thing, after all. From premise 2, it follows:

3. We can characterize all the "activity" of the universe in physical terms, at the bottommost level.

And now taking premises 1, 2, and 3 together, we reach the (abominable) conclusion:

> Everything that occupies a higher level, if it has an effect at all, must be having a physical effect—and that's where all the real action of the universe is at. The universe has a (micro) scale.

My view, to be elaborated in this book, is simply that premise 2 represents a gross misunderstanding of how things really interact—a misunderstanding fueled by philosophical neglect of the varieties of analysis in Physics. Premise 2 clings to a Cartesian mechanistic universe where all interactions are by direct contact. So the fundamental argument for the one-scale universe—resting as it does on premise 2—is ill-founded. From here on out, the goal of the book will be to show how premise 2 is false and leads to mischaracterizations of how the universe works. To remove the error, I propose we eliminate entirely the notion of level demarcations, since the notion of levels provides no useful philosophical ideas whatever.

Before we proceed, we need to do two things: (1) to characterize the way that sympathizers with this stripped-down argument will defend against my criticism and (2) discuss just how bad the argument's conclusion is.

First a remark about the argument's intended target. Its target is larger than the group comprised of those who, like me, are card-carrying levels-despising fanatics. Someone who resists reduction, even ontological reduction, but finds levels attractive, has to resist its conclusion—or anyway, might wish to. (The reasons for this resistance will be discussed anon.) What would a person have to do to resist that conclusion? The best strategy is perhaps to try to break the link between premises 2 and 3—ceding that entities interact with other things physically, but perhaps that we cannot characterize all the activity in the universe at this physical level.[2] It is very hard to make a case for this position. If all activity is physical action, then how is it possible that not all activity is characterizable in physical terms? The physical characterization would have to be incomplete. How is this possible if, as premise 2 suggests, "effect" *just means* "physical"? The effort will be ill-fated, though I will discuss it at some further length in chapter 1.

Now we should discuss the nature of the conclusion: that all activity takes place at the lowest level. Consider two objects roughly table-shaped, and going by the name of "desks." They are exactly the same in weight, size, and shape. Each of them is sitting in an identical, otherwise empty room on the second floor of a new house, which, unfortunately, one day falls victim to rising river waters. At the end of that calamitous day, both of these desks are found situated exactly one floor below their original positions, with gaping identical holes in the ceilings above each. How should we characterize what happened? The natural (and correct) way is to say, of each desk, that its *mass* or *weight* was originally *supported* by a platform whose integrity was at some point compromised by the water; the

[2] S. Mitchell (2009) seems to occupy this position: "If there is something in the world that can be isolated by the functional description (caused by X and causing Y), there is no reason to think that a physical description of that piece of the world, partial as it is, will be identical with a higher-level description of that piece of the world, partial as it is" (33).

result of this compromise is a loss of the support, which resulted in the final configuration of the desks.

How can such an explanation cause trouble for defenders of the one-scale hypothesis? There is one detail I neglected to mention: the desks have different physical constitutions: one is made of wood and the other is made of a petroleum composite. Still, my initial description of the desks is entirely correct; they weigh the same and occupy the same volumes. And they result in the same configurations in the unfortunate home. This case causes a problem for my opponents, however, because the account we gave in the paragraph above attributes action to our entities at the wrong level: masses or weights of desks are acting at a roughly human scale on floors. Masses and weights are mentioned, but each of these is *realized* by a different set of lower-level characteristics: in the one case by wood particles, and in the second by a polymer. (In general, mass and weight are realized by various different kinds of substances that are, if they are at all, held together in various and sundry ways. Still, differently composed things can share weight and mass; and much of Physics is done in terms of weight and mass, with no regard whatsoever for how these are realized.) But one-scale adherents should maintain that we have not *really* characterized the process at its most basic level. We need to go down further. And this would be all right, if indeed we could then find something *the same* about the two processes as we proceeded to a more micro analysis. The trouble is, however, that even if we could find a level at which the two processes looked identical, there is still no basis for the contention that this (second) sameness would be a better account of the sameness of fate suffered by the desks and their respective floors than the one that we just gave. There does not exist the minutest of reasons for supposing that the level of "mass" or "weight" is not really the right level for pitching the explanation. There is absolutely no reason to think that we shall indeed come abreast of *a second* level of description at which we will encounter a convincing sameness.

You might say: wait a minute, we already have a physical description—the one in terms of weight or mass. What's the problem? There is nothing left to do. The problem is that the case described is an instance of what is often termed *multiple realizability* of a property at one level with different properties at lower levels. Here we have a case of (physical) masses multiply realized by different lower-scale properties. This is uncontroversial. What is more controversial is what the diagnosis of multiple realizability portends. There are two options: either (I) multiple realizability is diagnostic of levels—in other words, multiple realizability is a mark of the existence of more than one level—or (II) it is not. Suppose (I) it is. Then this would force one of two conclusions: (1) the "physical" level resides even further down than the level at which we just described the bodies as possessing mass and weight, or (2) Physics occupies multiple levels (which might be a boon anyway, since we want a physical level at the cosmological scale). Each of these reveals problems with the notion of levels. In the first instance, we don't have any grasp of where the physical level "really" is. And in the second, we don't have the right characterization of the physical. For if "physical" occupies multiple levels, why suppose that Physics has any special role to play—for instance as an MS? There might be other reasons, but we don't have any hint of a suggestion of what they might be.

Suppose, on the other hand, that (II) multiple realizability is *not* diagnostic of levels. The realization of mass by lower-scale characters is not the mark of more than one level. Then what is the mark of a level? With that query, we are where I want us to be vis-à-vis levels: prepared to jettison them.

You might, ultimately, think that mere scale is the mark of a level. This is just as problematic as multiple realizability. Because, recall, we want cosmology to qualify as Physics. In which case, you give up the specialness of Physics. But then why hold to the specialness of the most micro? Where did anyone ever come up with a shred of an argument for that proposition? The only defense of it is that it makes for a kind of tidiness. The advantage of tidiness (and even if we set aside questions as to whether there truly is tidiness), such as it is, cannot compete with the advantage of being able to accommodate a wide swath of facts—such as the facts of interaction between the desks and floor, at that particularly macro level of granularity, among other facts that will emerge as we proceed.

6. There Are No Levels (in the Plural)

Something needs to be said about the notion of scale, by those (me, for one) who postulate the scale freedom of the universe. Scale is like the notion of dimension in the following regard: adding more of it (when some composite or aggregate of elemental entities exhibit multiple scales of activity) is not simply adding "more of the same." In other words, there are qualitative changes when there are additions of scale, just as there are qualitative changes when there are additions of dimension. If we lived in only two dimensions—if we were really Flatlanders, for instance—we would experience the world fundamentally differently. Likewise, adding activity at higher scales changes things qualitatively. So it cannot be simply that adding scale is adding inert material—so that we simply needn't take notice of the apparent activity at higher scales. (This analogy will become much clearer in chapter 7.) Returning to the example that started us off in this chapter, a new introduction to an old work can well make readers look at the work considerably differently than they might otherwise have done. (Think about the difference such an introduction would make if the work's original audience were in a position to read it!) These points, if you are prepared to accept them at this time, will prompt your rejection of my opponent's model rendered in figure 0.1. If you are not convinced yet, then I invite you to work carefully through the arguments of chapter 1.

Is there a right way to think about how scales interact? Is there a natural way to conceive of the difference that addition of multiple scales makes? To questions such as these I like to answer with an anecdote that is often told about Ludwig Wittgenstein.

> Wittgenstein's student (and literary executor) Elizabeth Anscombe once remarked "You can see why people thought the sun went around the earth." "Really? Why?" he probed. Anscombe replied, "Well, it looks that way." To which the teacher, still teaching, countered: "Well, and how would it look if the earth went around the sun?"[3]

When I say that there are no levels, I am not saying that somehow it was a good first try to postulate them, given the data we had at the time, and that now the data are in we see that the proposal was wrong. Rather, the evidence has been here all along that there are no levels. Physics itself all along has been producing scale-free accounts that tender counterintuitive (counter one-scale) predictions—which are robustly confirmed, as we will see. The invitation to entertain a scale-free universe is thus an invitation to a Copernican shift. What I hope I'm doing is helping philosophy get out of the way of the shift. Because, sadly, philosophical dogma (associated with attachment to the picture of figure 0.1) is all that stands in the way of progress on the metaphysics of the universe.

7. Can We Know the Scale of Our Universe?

You might well ask: how can we possibly know what the scale of our universe is? In other words, how would one go about determining whether there is genuine "activity" on a given scale? Don't we just have to take stabs in the dark, throw out guesses (call them *hypotheses*) and then test, as best as test can, to refute, confirm or at least corroborate a hypothesis? Of course. There is no other way. For obviously, in the beginning, the universe was much simpler than it is now (or anyway it is not crazy to suppose so)—and at that time we would not have required the kinds of pictures (or indeed the kinds of sciences) that show up in either of the figures 0.1 or 0.2 drawn in this chapter. But how does one test for the scales at which there is true activity, at this point in the history of the universe? One of the more important goals of this book is to *operationalize* this question. Here's a glimpse into how I hope to do that.

[3] The story is related in Smith (2008).

The question of this book can be asked another way: what are the degrees of freedom in our universe? On what scales do we see *independent* variables manifesting themselves?[4] This is the question that this book frames as important. (And the concepts that respond to the question are formulated most carefully in chapter 7.) This book also argues at great length, using cases from Physics itself, for the proposal that the degrees of freedom are not all micro. In fact the question of the identity of independent degrees cannot be known a priori, and so has to stay open permanently.

So long as the question is treated as an open question, so long as the question is not treated as foreclosed by a philosophical preference (howsoever sourced) for a one-scale universe, then there is a great deal of work to be done in metaphysics to understand the nature of objects. Understanding how objects "hang together" and how they change over a given interval of time (in those instances where in fact they do) requires at least in part making sense of the fact that certain features of systems are manifestations of its independent degrees of freedom—its *state leaders*, as I'll eventually call them—and certain others are manifestations of characteristics that *follow* the leaders in ways scripted by natural laws. Since we don't yet have an array of philosophical options through which to understand the possible ways that characteristics lead and follow, we require philosophical foundations for thinking about the question, before we can truly have viable theories of concrete objects. This is the other major contribution of the present book—by way reinventing a role for Philosophy in the space of contemporary sciences.

Once upon a time Philosophy played the role of an active coparticipant in the growth and development of Science; philosophers were at one and the same time scientists too, and did not insist upon an exclusive preserve that belonged only to Philosophy. In the twentieth century everything changed. Philosophy seems now to have vacated its original role as team player (captain, sometimes), and philosophers have for the most part turned away from the advancing frontiers of Science, having withdrawn to an exclusively philosophical cache of academic services, topics, and concepts. Of course it is quite possible that this should be regarded as progress in Philosophy itself—though I should not be among those who regard it so. Be that as it may, philosophical dogma such as is ossified in the image of figure 0.1 should not be allowed to stand in the way of other disciplines' understanding the universe. Whatever else philosophers might be allowed to do on their own turf (if they have one), ossified philosophical dogma

[4] Note that one must avoid the temptation to interpret "independent" here as "causally independent." By the end of chapter 4, it will be clear also that this "independent" is not identical with "uncorrelated with." For now, we may think of this as closer to the relationship in which the property of beauty stands to the eye of the beholder: the one cannot even exist without the other.

should not be allowed to hinder scientific progress. Sadly, that is happening. The task of this book is to remove that stumbling block.

8. The Importance of a Theory of Science

Articulating a metaphysics that allows us to posit the scale freedom of the universe is part of a theory of science. (By contrast, and quite unfortunately on account of the virulence or "stickiness" of the idea, it takes comparatively little by way of a theory of science to posit the existence of a master science or the reductive hypothesis of a single-scale universe.) So—as I gestured at the outset—we shall be required to do spadework in metaphysics, and in particular in the metaphysics of science, to defend the message of this book. Fortunately, the work will be worthwhile because it will be the foundation for a general theory of science, which will help to answer questions pertaining to the nature of science, as contrasted with other human enterprises. This question, for instance, is among the most famous philosophical problems of the twentieth century and is known as "the problem of demarcation": does it matter that everything called Science shares with everything else of the same name a "core" set of defining features? Or is it all right if there is merely a family resemblance between them, and nothing in general to signal the difference between science and other human undertakings?

Three major theories of science have been elaborated in the history of the discipline of philosophy: Aristotle's, Rudolf Carnap's, and Karl Popper's—in order of historical appearance. Each of these provides a perspective on the demarcation problem. I'll save discussion of Carnap's for last, though it is not historically last.

Aristotle's theory of science was the same as his theory of knowledge; the latter he understood as inquiry for its own sake. He held that inquiry falls into natural categories that today we might refer to as disciplines, and his primary contrast was that between knowledge (the product of inquiry for its own sake) and common sense (which serves many useful ends). Aristotle maintained that standards of inquiry vary across disciplines: for example that logic and mathematics have standards of proof that differ from standards of evidence in the study of nature, or in politics (which for Aristotle was also a discipline). He was clear that common sense is not a form of knowledge because common sense has no standards at all. It is not a discipline, but simply the result of a kind of practical activity of human life and so to be evaluated by its serviceability in those activities.

Aristotle's theory of science was positioned to answer the two most important questions in the area of (scientific) explanation: (1) how does a scientific explanation of something differ from a mere description or explication (or anything else) of it? and (2) what makes a scientific explanation actually scientific rather than something else? Aristotle's answers are (1) scientific inquiry results in an understanding of essences, first principles, and definitions of key (discipline-defining) concepts, that can be organized in syllogistic form, and which answer the question "Why?" And (2) scientific inquiry is governed by norms orienting it toward truth—norms that do not govern common sense. However, until we are presented with a clear sense of what the truth-orienting norms of science are, and how it is possible to fall afoul of them, any pursuit or community of seekers can claim to be in possession of truth-orienting standards—from phrenology to the various forms of divination—particularly if the so-called norms explicitly provide that the uninitiated have no access (legitimate or otherwise) to its true (and possible also actively shrouded) practice. Until philosophical light illuminates the norms associated with science proper, anything practiced in the shadows can raise itself up to an unwarranted height. This is the gap that Popper sought to fill in the early part of the twentieth century—and for which he has rightly been widely esteemed by the scientific community.

Popper made it his fundamental concern to demarcate science, as such, from things of lesser dignity (see Popper 1959). To this end he focused on the idea of scientific progress—that is, certain ideas could be advanced ahead of others that preceded them in time before the tribunal of science. The key idea here is that being first in line, when it comes to scientific proposals, should carry no weight in their final reckoning as to truth. Like many of his contemporaries, Popper took note of the fact that observation alone could not advance a hypothesis ahead of its competitors, since any given body of observations will be consistent with numerous and mutually exclusive bodies of theory. He brilliantly maintained that the core of a philosophy of science would account for how an idea, however initially outrageous, can grow in scientific stature, to the point of eclipsing or even displacing older and perhaps even initially better-championed theories. This account would spell out the method of scientific comparison, of the testing of one eligible hypothesis against another. For the distinction of "scientific method," Popper proposed the method of falsification. According to this account, a theory is scientific to the extent that it excludes or prohibits certain possibilities that are in principle observable—and is *ipso facto* falsifiable; conversely, it is corroborated to the extent that it survives tests aimed at falsifying it. Popper's critics were later to argue that the method does little in the way of providing a theory of comparisons—it is merely a measure of the severest test of a hypothesis, and not against other theories but against nature itself. Rarely are the conditions for such a severe test available, because hypotheses can be shielded from falsification in numerous ways (that we need not discuss here). The point to stress is simply that Popper's contributions occasioned the philosophical community to mark an important distinction vis-à-vis the functions of scientific theory: in the first instance, scientific hypotheses receive attention as objects of scrutiny—they are subjected to tests of *confirmation*; however, once so scrutinized (and surviving scrutiny), scientific hypotheses might subsequently be called on to perform a range of other services to which the label of *explanation* might be appropriately attached. Not everything merely eligible of scrutiny (falsifiable) is called on to perform the task of explanation. Theories of explanation take up the subject of what Popper leaves largely unelucidated in Aristotle's theory of science.

Finally, the notion of a Master Science originated in the early twentieth century, within the Vienna Circle of scientifically trained European philosophers, with a zeal for science, its status, its language, and general principles of conceptual analysis. These philosophers were first to articulate the notion of the unity of science, and to use the notion of "levels" in this articulation. The model of unity they chose, first developed by Auguste Compte, culminated in Carnap's *Logical Structure of the World* (1928). In Carnap's magnum opus, the sciences are unified through the unity of their subject matters (ultimately constituting one real world), which is evident in the relations between their concepts. Roughly, scientific terminology makes reference to the world, perhaps even to the same regions of it, but different sciences make contact with the world at different "levels" or "strata." This fact corresponds to the relations in which stand the conceptual apparatuses of different sciences: the concepts of the sciences stand in hierarchical relations, corresponding to the hierarchy of the sciences themselves, and this hierarchy is organized roughly along the lines of scale and complexity. Higher levels, on this account, result from sedimentation, rather than from true activity.

C. D. Broad, a critic of the Vienna Circle program and typically counted among the first emergentists, was later to characterize the foundation of this unity as follows:

> [There] is one and only one kind of material. Each particle of this obeys one elementary law of behaviour, and continues to do so no matter how complex may be the collection of particles of which it is a constituent. There is one uniform law of composition, connecting the behaviour of groups of these particles as wholes with the behaviour which each would show in isolation and with the structure of the group. All the apparently different kinds of stuff are just differently arranged groups of different numbers of the one kind of elementary particle; and all the apparently peculiar laws of behaviour are simply special cases which could be deduced in theory from the structure of the whole under consideration, the one elementary law of behaviour for isolated particles, and the one universal law of composition. On such a view the external

> world has the greatest amount of unity which is conceivable. There is really only one Science, and the various "special sciences" are just particular cases of it. (1925, 76)

Broad recognized this as more than physicalism—as more than a preference for the substantial and observable over the private or "spiritual." It is fortified with a preference for Physics—with a dogma of Physics as Master. Broad rejected it accordingly.

Defenders of the master science thesis might naively respond that it is just a matter of logic or mathematics that physics comes first. How could it be otherwise? Doesn't the physical manifestation, the physical aspect of the universe, have to come first in time? Of course this reply presupposes what requires argument: that something must occupy a special place, a place of honor, in the roll of sciences, because its subject matter must in some sense come "first." I will be responding to this point and others much like it throughout this book. Here I will ask the reader simply to bear in mind that members of even the antireductionist leagues—those who reside in emergentist camps, for instance, as we shall see in chapter 1—are not as thoroughgoing in their repudiation of the Master Science notion as Broad was. Some embrace it, and accordingly do homage to Physics as the Master Science.

Emergentists in today's intellectual climate, some of whom aim to defend autonomous sciences, proclaim that the unity of science does not lie in the reducibility of the various sciences to Physics, but rather in metaphysical relations of the entities and properties recognized by "special sciences" to those recognized by Physics. Thus emergentism in contemporary hands still commits the error of acknowledging a Master Science, but does it as a matter of metaphysics: it comes in the form of a master ontology of independent entities.

Carnap's conception of unity was never very popular among practicing scientists, who in the preponderance found Popper's falsifiability criterion much more satisfying as a definition of their field, or at any rate as a definition of the norms of proceeding in their field. Nowadays scientists (and philosophers too) seem much less agitated about the problem of demarcation—or perhaps they are content to defer to Popper's solution.

Still, there is something distinctly attractive in Carnap's idea: it locates the unity of science in the subject matters of the sciences. Unfortunately, the model requires an intermediate between science and the world: language or at any rate some system of concepts. My model (displayed in figure 0.2) simply does away with the intermediary, while at the same time removing levels. Everything else is intact. Most importantly it preserves unity in the fact of overlap between the subject matters of the sciences. So it preserves what is attractive in Carnap's model. That is in its favor.

Finally, you might be concerned that my model does not adhere to the Vienna Circle's deep concern: materialism. This concern is in some regards disingenuous. Physics, at least since the Vienna Circle, is comparatively unrestrained when it comes to positing wild and crazy kinds of entities and processes. Just consider for a moment the proposals of dark matter, dark energy, things that pass through us in enormous numbers without leaving a trace—to say nothing at all of alternate universes and cats in wildly unstable states between life and death! And nobody freaks out—not in a censorious way. Nobody threatens to call the science police. True, there is usually some empirical provocation for the postulations, but nothing like definitive evidence. And this is as it should be. False hypotheses, so long as they're testable, should always be welcome. No crazy nonsense should qualify as science (a quality control no one can object to), but one can't know what is crazy nonsense in advance (remember the Wittgenstein parable).

I stand with those who say there are no ghosts, no demons, no ectoplasm—in this regard I'm on the side of the angels (fittingly, or ironically, as might suit your taste). But in accordance with the stance of this book, this position of mine on ghosts and ectoplasm can have no normative traction against yours, if you wish to consider their existence seriously. I have no legitimate reason to debar you from science if you disagree with me. One of the most creative thinkers in the history of quantum thought—the father of quantum computing, foreseer of superfluidity, and Nobel laureate to boot—has been willing to stand (scandalously, as some would insist) on the side of the paranormal. I'm not about to join Brian Josephson, but we'd be doing the scientific community no favors by branding him a kook and removing him from our company, because he might in the end be right.

9. The Master Narrative of This Book: The Unity of Science

Here, in epigrammatic form, is the thesis of the present book: there are no hierarchies in nature, and so the sciences require none. There are enclaves of order, to be sure, corresponding to biology, psychology, economics, and what have you. Discipline names mark out schemes or regimes of order. And these are for the most part relatively independent of one another, insofar as the laws of order in each scheme are relatively independent of (most) elements or properties ordered by the laws or structures prevailing in another scheme. So to some extent the subject matters of each discipline can be studied independently of the others—and so are autonomous. But I will be arguing that this is no basis for positing a hierarchy of disciplines, much less a hierarchy of entities or characteristics. I will argue that hierarchies in general are downright antiscientific, downright inimical to the spirit of free and open inquiry. They suppress profound questions.

But this is not a book with a purely negative message. It has a positive one too. It provides an alternative metaphysic for organizing the universe. This is already previewed in figure 0.2 and will be further explicated in the rest of the book. Indeed, the depiction of unity in figure 0.2 prefigures the master narrative of the book: I am affirming the unity of science, but a unity without hierarchies.

The book comes with a positive program of research as well—which starts with arguments for the positive impact of figure 0.2 as well as a metaphysics for making sense of it. Indeed the figure is fertile with questions for exploration—radical questions having to do with the natural history of order in the universe. The picture thus promotes close encounters with these questions of the natural history of the universe, questions about major transitions in its organizational evolution. How are natural features on different size scales related to one another? More generally, what are the types of relations in which features (whatever their scale) might enter into? These are very general metaphysical questions about relations between the characters that show up in our world—"quantities," as scientists like to call them.

There is no one universally accepted taxonomy of such relations. Indeed there is no taxonomy of quantities themselves, save by reference to scientific discipline. So, for instance, a physical property is routinely construed simply as one that enters into physical theorizing (theorizing done by physicists), in an especially proprietary way. This is of course immediately problematic, as properties that attract the attention of physicists—such as spatial separation, temperature, and many others—tend also to attract scientific attention elsewhere as well, and feature in treatments of a very wide and diverse range of phenomena outside the ranges usually construed as belonging to Physics. When a property enters into a scientific treatment, there is (invariably) no unambiguous signal that the property in question is being treated in a proprietary fashion. This being the case, it will not do simply to gesture blindly at disciplines in hopes that boundaries between disciplines can go proxy for boundaries between taxa of quantities.

The question immediately arises whether we can be in a position to postulate anything like a taxonomy for the class of properties in advance of empirical inquiry—or at any rate, in advance of inquiry about the scientific disciplines themselves. And so what serves best, at least initially, is a taxonomy of ways in which—or a class of models of how—that class *might* be structured. The discipline of metaphysics is best served by a clear and rich taxonomy of models of possible structure of the class of properties to guide subsequent inquiries. (Subsequently, the logical next step—though this is still a very long way off—would be to articulate how potential empirical findings would favor or disfavor each of the models to be put on the table.)

As a number of authors suggest, quite reasonably, it might actually help in the effort to identify potential structure within the category of properties if we were to examine the findings, or at any rate the methodologies and strategies on display, in a variety of scientific disciplines.[5] So, for example, if we look just at strategies for handling relations among properties within the discipline of Physics itself, we notice the following. In Physics there are three types of relations one can identify among physical variables. The most familiar to the college student is the ubiquitous functional or covariational/correlational relation found in any equation. The second, and less familiar, are the relations between whole categories of properties illuminated on the occasion of a derivation of a limit theorem of one sort or another. (An example of that sort is as follows: *relativistic momentum* and *classical momentum* are equal in the classical limit as $\frac{v^2}{c^2} \rightarrow 1$.) The final and by far least familiar kind of relation is reached via the mathematical procedure of renormalization. I discuss this procedure at length in chapter 8.

Now, functional relations can be interpreted either as strict property identities or as law-like regularities. When the former occurs, a reduction is declared, or at any rate a discovery that two names have been given to the very same property—corresponding perhaps to two distinct measurement procedures of it. Limit relations are subject to both readings as well. Relations illuminated through the process of renormalization, by contrast, are almost uniformly ignored by philosophers. Some authors look to limit relations as sources of structure in the class of properties, arguing in favor of either one or the other principle by way of structuring the class into levels.[6] I will be arguing in the chapters immediately to come that there are no good arguments for levels. Furthermore these principles for structuring the class of properties into levels cannot be the basis of a natural history of these domains—indeed postulating a hierarchy stops from answering questions about their natural history. The remainder of this introduction will be devoted to this important argument.

Consider the domain comprising the phenomena of biology (including micro and molecular phenomena) and ecology. And consider also all the organizational principles, laws if you like, that direct these specifically biological patternings.[7] Similarly, consider also the domain comprising phenomena of the *mental*—or, if

[5] For example see Batterman (1995), and Rueger (2000).

[6] Batterman (1995) is perhaps the best instance of the former and Rueger (2000) of the latter. A good overview of current formulations can be found in J. Wilson (2005), who also argues at length for the inadequacy of supervenience.

[7] Fastidiousness or high-mindedness about referring to the patternings or the principles that produce them, as *laws*, on grounds that historical sciences cannot really exhibit full-fledged laws, will not be an issue here; the use of the term "laws" here is purely for expository purposes. The concern with patterns, as we encounter it in this book, revolves entirely around the origins of these things and their explanational locations, in the larger order of explanatory schema, and not specifically their status as laws.

you prefer, simply the *cognitive.* A nonreductive physicalism these days will insist that in each case, the subject matter (the cognitive or mental, on the one hand, and the biological and ecological, on the other) is in an important way distinctive from that of Physics. The organizing principles of Biology and Psychology are, in some sense, independent of Physics.

I want to affirm this as the fundamental and most defensible impulse behind a nonreductive model of the domains of diverse sciences. In my opponents' language, it is because new "layers" of organization are emerging all the time, there has to be something genuinely "new" coming to be (something "emergent," as for example Samuel Alexander, whom we meet in the next chapter, will have it). Now consider this question: how did the phenomena of Biology or Psychology originate? What explains the putative independence of these phenomena from the phenomena of Physics, as well as the independence of their explanatory principles? I think that this question has an answer—a thoroughgoing empirical one. The answer is not a purely metaphysical one—the purely metaphysical one being simply that this independence exists, and that it amounts to some particular specification of conditions that render the two domains independent on the basis of some broadly "logical" features of the properties in question. I think that the question concerns the natural history of systems of patternings among quantities. How do patterns of covariation among quantities emerge? And how do they manage to cluster so as to create enclaves of order?

More to the point here: if there is a natural history of order, where in the hierarchy (if there were a hierarchy) would it belong? Could it belong to the arena of Physics? Hardly, since that arena recognizes no laws or patternings in which (for example) biological traits feature, as such, or for that matter novelty of any other kind. (This is because, according to the precepts of those who maintain the hierarchy, the events at higher levels are "decoded" at levels below—they are articulated in physical rather than higher-level terms.) Indeed if we could explain their emergence with purely physical explanatory tools, this would appear to be a self-defeating demonstration—a reductio ad absurdum of the independence of biological phenomena specifically, and so nonreductive physicalism generally.

Could such an explanation, by contrast, belong to Biology or Psychology, or to some other elevation in the hierarchy? No, since a biological or psychological framework would have to presuppose its own legitimacy as a framework for explaining these origins, and so would hardly count as an explanation of the emergence of biological organization. We might suppose, as a working hypothesis, that there is another level, perhaps a "meta" or penthouse level, at which these origins questions can all be addressed. But this is problematic. For it is not just one fact, this question of origins, but many—one for each new level. Each one positively empirical. If we add just one penthouse level for all, do we not owe our constituents an answer as to how *this* level itself originated, if not why it deserves to be called a new level, at penthouse elevation? Its appearance is purely ad hoc in this schema—a way of putting off an embarrassing question.

To summarize the argument thus far: Physics cannot tell us why Biology exists—why Biology as an "autonomous discipline" is even possible—and Biology cannot explain its own emergence without taking its explanatory framework as given and therefore immune to explanation. So, if we assume the existence of a hierarchy, in which each phenomenon of nature must be slotted, the question we are now considering will be unanswerable as a matter of natural fact. It will be "purely metaphysical" in the derogatory sense, deserving no serious answer: explaining the emergence of organizational levels, the major transitions in organizational evolution, as a question of natural history will have to be relegated, as a matter of necessity in the order of things, to the discipline of pure metaphysics—a place it does not belong. There will be no scientific place to inquire about such large questions. So hierarchy stands in the way of scientific inquiry.

These considerations, to my mind, serve as one large *reductio* of the hierarchy as a purely metaphysical edifice. We require a true account of the autonomy of the sciences. Since the doctrine of levels interferes, we are better served by its demise.

Repudiation of the idea that the physical (meaning the micro) enjoys ontological priority might seem like a threat to science. I've sketched here already some reasons for thinking that, on the contrary, it is a good thing for science. More reasons will surface as we proceed in the book. It will become quite clear that the levels-rescinding model I propose does not invite dualism or other antimaterialist doctrines—to the contrary, it demands that characters be admitted into the ontology of science only as we come to understand their operations within an entire body of scientific principles and models. Clinching the case for dropping levels will require providing an alternative model of the unity of science, which makes explicit strictures against antimaterialism unnecessary. This is the scale-free model. In chapter 7 I will argue that this model is consistent with materialism and is nowise a repudiation of it.

This book is essentially in three parts. The first, including chapters 1 and 2 and this introduction, provides an orientation toward the project—an introduction of the main thesis of scale freedom, some discussion of predecessors and their origins, and a variety of contextualizations for different readerships as to why we need a scale-free model of the universe. Part II then proceeds with the handling of foundational philosophical questions (like fundamentality, causation, the progress of science, the cement of the universe), suitable for a theory of science compatible with the scale freedom of the universe and truth. Part III further illuminates these arguments using Physics as a microcosm of science more broadly construed. In chapter 7 the concepts of dependence among quantities is presented, as well as a more graphically oriented argument for dispensing with levels and level boundaries. In chapter 8 we discover that Physics has really already embraced the scale-free universe, and that the theory of renormalization has resources to articulate this scale freedom—but that at the moment all this is shrouded in a language of "paradox." We'd do better to remove these shrouds that veil the message that the universe is truly scale free.

PART ONE

SCALE FREEDOM MULTIPLY CONTEXTUALIZED

1

Against the Philosophy of a One-Scale Universe

You can't teach an old dogma new tricks.
—Dorothy Parker, quoted in *The Algonquin Wits* (1968)

1. Introduction

Our universe is scale free. By this I mean that is active at multiple scales, potentially at every scale. It's not true that all the "real" action happens at the smallest size scales; it's false that the large-scale features are simply inert, and that all of us—large and slow-moving beasts that we are—are banished from the true dance of the universe. The one-scale model of activity in the universe is a dogma too old for new tricks. This book is devoted to building a credible case in favor of a model that inverts the old dogma, and this is meant as a way of making a beginning at building a credible case *against* the one-scale model as the philosophical conception of the universe we actually inhabit. The one-scale model is the alter ego of the hierarchical model of the sciences, uncompromising in its hierarchies. Since they are two sides of the same coin—the one in ontology and the other in the theory of science—I will use the terms "hierarchical model of the sciences" and "one-scale model of the universe" interchangeably.

The scale-free universe model *inverts*—or perhaps one should instead say *subverts*—that hierarchy as well, since its alter ego is a hierarchy-free theory of science. Fittingly, then, we shall be rejecting a one-scale, many-levels model and replacing it with a one-level, many-scales model. To be clear: "scale" refers to size, and the scale-free model affirms differences of scale—it acknowledges that there are such differences, and that one needs to take account of scale differences to sort out what is going on in the world. However, the scale-free model insists that differences in scale do not amount to differences in "levels"; distinctions as between levels are, as I shall argue, philosophically unworthy.

The next several chapters will present a kind of philosophical revue—a series of interludes intended to clear away mistaken preconceptions that prop up the hierarchy, and more philosophical interludes to set metaphysics on a truer course toward scale-free modeling of the universe. Most of the passages you will encounter here work through examples—a large diet of them. Many of these will recur throughout the book.

Before discussion of example cases can have its intended impact, we need to make more explicit, in as precise philosophical terms as possible, the one-scale universe model, utilizing the terms of art that philosophical researchers employ. In the interests of full disclosure (and as you might already have suspected) I will ultimately propose dispensing with this terminology in favor of language that does justice to a scale-free universe. So while this exercise will be one in clarification for persons steeped in current philosophical usages, I believe the terms in present philosophical currency can be disorienting to those who are already walking in the light. So the reader unfamiliar with current philosophical language can feel free to pass through certain passages of the present chapter—and I will mark them in advance—lightly or not at all. The exercise in the present chapter is a bit like rubbing your pet's nose in the mess it made, so that it will not do it again—a form of aversion therapy for which Wittgenstein is rightly credited, and in which philosophers quite generally indulge in the present era. But, to repeat, the exercise can be somewhat disorienting to an individual unacquainted even with the furniture, let alone the mess. Still, it is a cautionary tale, and at least in that respect a lesson of a certain kind.

Our journey begins where the philosophical community is currently mired. But we will leave much of the mire behind by the conclusion of this chapter. This chapter will thus be at least in part a first contextualization, meant to orient contemporary philosophers toward the project, as well as to introduce some of the important examples from Physics that will be staple throughout the book, while also acquainting the wider readership with relevant philosophical terminology. So begins another contextualization.

2. Emergentism

A reader alive to contemporary philosophical literature is sure to suggest that the message of scale freedom that I am trying to foist upon innocent would-be consumers of specialist scholarship already exists and bears a rather elegant name, with distinguished ancestry—that it is already known by the label of *emergence* or *emergentism*. There are now many versions of emergentism, with some forms allegedly being "stronger" and others "weaker," some palatable to the reductionist I described in the last chapter, and others not so much. An alert reader might suggest that my view—by the inelegant name of the scale-free universe proposal—might be a very strong, perhaps even radical, form of emergentism, but a version of it all the same. After all, both are apparently in pursuit of an articulation of the (admittedly vague) credo that the whole is in some sense greater than, or transcends, the sum of its parts. At any rate, that is the emergentist's ambition.

It might be that (as that alert reader suggests) some defenders of emergentism will approve of my proposal—perhaps some of those on the more "radical" end of the spectrum. Notice first, however, that the way I have articulated the issues makes it altogether transparent that there are no clear in-betweens in between a one-scale universe and a scale-free one. In answer to the question of the scale of "true action" in the universe, one can take only one of two positions: that the universe has a single fundamental scale, or that it has none. The latter is my position. So, vis-à-vis scale freedom, one is either all or nothing; one cannot be more or less radical than anyone else on the issue. Even if (hypothethically) one were sympathetic to the idea that the universe is active at just *two* scales, so long as they are distinct, that is still a scale-free universe because there is no unique single scale. So there is no room for in-betweens on my articulation of the options. How can anyone be more radical on the issue than I am? Or less, for that matter, and so more moderate? Further, there is no "emerging" on my view. True, there might have been eras in the life of the universe where there were fewer scales—or indeed more—at which there was real action than there are today. But it's not as if action at higher scales *keeps* emerging from below in a sustaining way, as the emergentists imagine. My articulation of the issue must therefore turn on something rather different from what the typical emergentist has in mind—and this chapter is meant to demonstrate this difference. Still, it might be worthwhile examining with some care how the advocacy of emergentism has proceeded, so that we might see how the position for which I am about to launch a case might benefit from scrutinizing its failures.

Emergentism reached its pinnacle in the British Isles in the first quarter of the twentieth century. But its roots go back further still to John Stuart Mill, who was writing somewhat earlier. In his highly systematic approach to the philosophical foundations of science, Mill (1874) was careful to note differences between the sciences, and he focused on the differences between physics and chemistry:

> I shall give the name of the Composition of Causes to the principle which is exemplified in all cases in which the joint effect of several causes is identical with the sum of their separate effects.
>
> This principle, however, by no means prevails in all departments of the field of nature. The chemical combination of two substances produces, as

> is well known, a third substance with properties entirely different from those of either of the two substances separately, or of both of them taken together. Not a trace of the properties of hydrogen or of oxygen is observable in those of their compound, water. The taste of sugar of lead is not the sum of the tastes of its component elements, acetic acid and lead or its oxide; nor is the color of green vitriol a mixture of the colors of sulphuric acid and copper. This explains why mechanics is a deductive or demonstrative science, and chemistry not. In the one, we can compute the effects of combinations of causes, whether real or hypothetical, from the laws which we know to govern those causes when acting separately; because they continue to observe the same laws when in combination which they observed when separate: whatever would have happened in consequence of each cause taken by itself, happens when they are together, and we have only to cast up the results. Not so in the phenomena which are the peculiar subject of the science of chemistry. There, most of the uniformities to which the causes conformed when separate, cease altogether when they are conjoined; and we are not, at least in the present state of our knowledge, able to foresee what result will follow from any new combination, until we have tried it by specific experiment (Mill 1874, bk. 3, ch. 6, sec. 1).

This did not lead Mill to disparage a rightful place for chemistry among the sciences, as an independent discipline in its own right. Indeed he wrote specifically, in the paragraph immediately following this passage, that "there is no reason to despair of ultimately raising chemistry . . . to the condition of deductive sciences." All it takes is articulation of the appropriate "laws," as he called them—and thought they were structurally of a different but hitherto unknown basic form—by which the phenomena they describe operate.

Mill's optimistic attitude toward the discipline of chemistry is not nowadays shared among physicists, and it was not even in the early part of the twentieth century.[1] Herbert Dingle, for instance, infamously proclaimed:

> [T]he truth is that chemistry indeed has no place in the strict scientific scheme, and that this is so can be seen from the fact already evident even at the present stage of scientific progress, that the ultimate generalizations of chemistry are all derivable, and indeed must inevitably have been reached sooner or later, from the development of physics itself—chiefly the departments of optics and electromagnetism. The periodic table, originally a product of chemical research, is a product also of spectroscopic

[1] J. van Brakel (2000, 15) attributes to Mill—mistakenly, to my mind—the view that chemistry is a failed science. I have quoted Mill at length to allow him to speak better for himself.

> research, and with this difference, that instead of showing in each place a chemical symbol and an atomic weight, the spectroscopic table shows a configuration of electric charges which, when fully, understood, will undoubtedly prescribe all the varieties of chemical combination that are possible. The whole of chemistry may therefore, so far as final results go, be regarded as a superfluous study. (Dingle 1952, 210)

Dingle's position echoes the oft-quoted remark of P. A. M. Dirac (1929):

> The fundamental laws necessary for the mathematical treatment of a large part of physics and the whole of chemistry are thus completely known, and the difficulty lies only in the fact that application of these laws leads to equations that are too complex to be solved.

It resonates too with the model of unity in science à la Carnap that I discussed in the introduction, whereas Mill's more cautious position is closer to mine. Mill is often classified as an emergentist. But to the extent that he refused to assign a priority to Physics, his dissent from emergentism is broadly in line with mine. He writes:

> All organised bodies are composed of parts, similar to those composing inorganic nature, and which have even themselves existed in an inorganic state; but the phenomena of life, which result from the juxtaposition of those parts in a certain manner, bear no analogy to any of the effects which would be produced by the action of the component substances considered as mere physical agents. To whatever degree we might imagine our knowledge of the properties of the several ingredients of a living body to be extended and perfected, it is certain that no mere summing up of the separate actions of those elements will ever amount to the action of the living body itself. (1874, bk. 3, ch. 6, sec. 1)

For Mill, the processes of life owe little if anything to Physics construed as the reality of inorganic compositions. And in his assigning Chemistry a rightful place in the roll of sciences, he gives a nod, in my mind, to the notion that there is activity at every scale of measurement—in essence, to the scale-free universe.

Samuel Alexander (1920) had a rather distinctive, and I think more evocative, way of putting the general point around which emergentists today coalesce. His point (made in this passage in relation to the physical "basis" of psychological states) is that certain complex physical states are special:

> We are forced, therefore, to go beyond the mere correlation of the mental with these neural processes and to identify them. There is but one

> process which, being of a specific complexity, has the quality of consciousness. (vol. 2, p. 5)

He thought that certain physical states are "something new, a fresh creation" (7) involving a "distinctive quality" (55) that *emerges,* rather than merely *results,* from the neural process (14).

> The [emergent] quality and the constellation to which it belongs are at once new *and expressible without residue in terms of the processes proper to the level from which they emerge.* (45, emphasis added; cf. 67)

Contemporary emergentists are perhaps best characterized as wishing to moderate the kind of position occupied by Dirac and Dingle, to make room for the nonphysical as legitimate, but also to curb excesses of ontology so as to maintain undiluted their adherence to physicalism—a concern that in no way governed Mill's reasoning on the subject. Today's emergentisms are species of so-called *nonreductive physicalism,* while still (so very unfortunately) wedded to the notion of levels or strata of reality. The idea is this: everything that exists concretely (as contrasted with things that are nonconcreta, like numbers and other mathematical abstracta) is *ultimately physical*—made up of physical stuffs, and governed by nothing but physical laws. While there is a very wide variety of ideas that get mixed and matched by way of trying to make sense of this idea of "ultimately physical," I will here list four of the most basic ones and discuss them in relation to my scale-free universe hypothesis.

Here are some of the main ingredients that go into an emergentist brew. Some brews, one should note, omit one or more of these; and formulations of emergentism stress certain ingredients more strongly than others:

1. *Physicalism.* Everything is made up of the smallest units of physical/material stuff.
2. *Novelty organized by levels.* Things are organized in levels, with some stuff up high that is in some sense "new" vis-à-vis that below it; and there's no guarantee of prediction or explanation (no reductive relation of any kind) of higher-level characters or traits in terms of those found on lower levels.
3. *Supervenience.* There's no information to be got in terms of higher-level characters or traits—no fact statable in terms of higher-level characters—that isn't already "present" in those at the lowest level.[2]

[2] The typical slogan is "no difference at supervenient levels without a difference at the base level." Thus the equivalence: coding for a difference at a supervenient level requires different information at the base level. And hence this is tantamount to the idea that there is a dependence relation between the two levels of facts—though it does not clinch that the difference is one-way, as we will discuss later.

4. *Downward causation.* Entities and characteristics featured above can act causally upon those below. (This is the language typically used in the philosophical literature, but in terms utilized in this book, we would say that there is activity at every scale.)

Three preliminary remarks before proceeding into the lion's den. First, from what I've already set out forthrightly, it is abundantly clear that I must agree entirely with the principle of downward causation (4)—except of course that since I don't believe in the organization of things by "levels," there can be no "downward" to my "causation." (And in any case it's not "causation" either, but what's in a label? That is a matter for chapter 4.) It's worthwhile mentioning that a cottage industry has grown up around the effort to do away with the notion of downward causation as a central element of emergentism, but to my mind quite unsuccessfully (see also Greetis, in progress).

Second remark: I have already confessed at the end of the introduction that I too believe that everything is physical and that is that. So there is just that one level as far as I'm concerned, if it's worthy of the name when it's all by its lonesome. Still, if something or some feature exists concretely, it's simply, unproblematically, and indefatigably physical. And just to be clear, numbers and other mathematical, logical, and conceptual entities are not concreta, and so aren't in the universe of things we are talking about—the universe of the encounterable in space and time.

I am of a mind that it's really not worth bothering with stating something so obvious as this physicalism to which I adhere. The reason is that, in my conception of things, the term "physical" has no contraries—nothing that could possibly exist qualifies as nonphysical except insofar as its very description ("ghostly," "nonphysical ectoplasm," etc.) would qualify it as such. And of course such a qualification would immediately mark it as *not* among those things one might encounter, not of the actual world—and so positively nonexistent. This has the result that I can view each of (1) and (4) as simply trivially true, or else as saying the something substantive that I wish to affirm as well. So as to the first- and last-numbered of these four principles, emergentism is by my lights on the side of the angels.

The third remark is this: again obviously, I must explicitly deny (2) because I explicitly deny any plurality of levels. Still, I have to confess that when emergentists *name* some of the features they find novel at what *they* call the "nonphysical" levels—maybe they name a certain chemical or biological trait or character—I have to agree with them that these features are not reducible to (not explained by, not predicted by, and in every other way not *due to*) the ones that *they* name as physical. So (2) seems innocent enough, again *modulo* the nomenclature about levels. (Not every chemical or biological character will be deserving of this status, so like the emergentist, I believe we require a way of categorizing higher-scale traits. And I opt simply for the label "higher-scale traits.")

But do I really agree with the emergentists about this novelty business? What does the emergentist typically mean by novelty? At least two things, and these two things are negative characterizations of the novelty concept, not positive. First, the emergentist will explain that the relevantly novel feature is not something that is itself a feature of the parts. So, for instance, each of the numerous parts of desks such as we encountered in the introduction have mass that can be assessed and ascribed to them, separately from all the other parts, so that the mass of one of these desks taken entire is equal to the sum of the masses of its parts taken separately. Mass-at-desk-level therefore is not new, relative to its parts. So mass, for the emergentist, isn't novel. I agree: there's nothing new in the mass of the whole that isn't already present when we tote up the sum of the masses of the parts—when this purportedly a priori fact is actually true. Second, the feature in question is not a simple feature of the *aggregate* comprised of the parts taken together just so in the relevant ratios and relevant arrangements. So the desk-shape of desk parts, when they are all put in their proper places according to the manufacturer, doesn't belong to any of the parts separately, but desk-shape is not novel because it's merely structural. And again: I agree. These kinds of features of things, according to the emergentist, is what "novel" *can't refer to.* But it seems as though anything *else* is fair game, as far as novelty is concerned. A trait that passes the test that is *not* "summative" of a parallel feature exhibited by the parts, and also *not* structural, is eligible for the label *novel.*

Why does it matter that the relevant features be "new"? What work is the concept doing in the analysis? Well, the emergentist has to have some *means* of identifying the features of the world that are in some sense "emergent." (Note that I by contrast don't have to identify "novelty": I simply say that there is activity at every scale—it's not confined simply to the micro. I refer to size scales, whereas they wish to refer to something else, something more subtle, something more metaphysical. Emergentists don't care about activity so much as they care about *novelty* at each scale—or rather, "level.") The important thing for these philosophers is that there be a group of features, at each level, that reaches a new form of reality—a reality independent of ("above and beyond" as they say) the realities of constituent elements. So for instance one might suppose that individual features *fuse* (this term is due to Humphreys 1997a, 1997b) together to create a novel feature that somehow does not exist when the entities that compose the aggregate are not interacting. From that time forward there is a new higher-level entity with novel causal powers, and the constituents are effectively disappeared from it. Alternatively, one may suppose that description of the interactions of constituents of an aggregate, over a period of time (noninstantaneously considered), has to be chunked accordingly—that is, one cannot attribute higher-order properties without referring simultaneously to noninstantaneous microfeatures of systems. There are higher-order properties, all right, but they emerge as a result of over-time interactions between microconstituents. Static snapshots don't contain enough information to render higher-level properties visible. (O'Connor 2000 maintains this stance, and it's less explicit in S. Mitchell 2009.) To put it more technically: the higher-order properties *do not supervene* upon instantaneous lower-level traits, but they *do supervene* (on at least some potential emergentist accounts) on a cross-section of over-time lower-level traits.

This has the consequence that two aggregates, with all their proper parts being molecule-for-molecule identical and configured in exactly the same way, could nonetheless differ in their higher-level features. One might, for instance, be a live fruit fly, but the other one not so much. Does this proposition strike you as absurd? If it does, you are precisely the sort of reader I need to convince—the one who would gravitate to the supervenience principle that I will say must go. You are, in other words, the sort of reader who finds the principle that macro supervenes on micro very attractive. So you are the kind of reader that the philosophical community takes very seriously, and the one to whose anticipated criticisms they are most sensitive. As for the other kind of reader—the kind of emergentist prepared to deny the supervenience principle—can that reader be adhering to any kind of emergentism at all? For if one says that the higher-level trait is absent in one case, but not in the other, in what sense does the trait "emerge" from its micro basis? If the connection between the higher level and the lower level is not to be relied upon in this synchronic way, what reason could there be to insist on its being "emergent" rather than merely law-like related (as any micro property might be)? Rather than being special, these become extremely ordinary. (This complaint is due to Jaegwon Kim, who will be prominent on our stage momentarily. My advice to the targets of this criticism will be this: simply drop the "emergentism" from your vocabulary, and join the "scale-free" chorus.)

Now for my part, I have nothing against "higher-level" being ordinary, old, and nonemergent. And, as already indicated, I too acknowledge a distinction between those larger-scale characteristics that are merely "summative" or "structural" and those that are not. But I don't think that characteristics that fall into this category are in any way transcending a level or crossing a metaphysical boundary, so long as we recognize they nonetheless merit attention over and above that devoted to the composing entities and their features and dynamical behaviors. So it seems that I am defending emergentism, even radical emergentism. "New" in my lexicon just means "different." So you might—with the original alert reader—think that the difference between what I am saying and what the emergentists propose is only a difference around the edges—that I simply have some idiosyncratic issues with the "levels" language. So what if they organize things in levels and I don't? I shall argue there is a gargantuan difference. The levels are, as I propose to convince you, a true abomination. They are surreptitiously bringing together the radical emergentists (whom I can defend when I push aside their talk of "levels") with the "moderate" ones—the ones who adheres to supervenience of everything upon the "microphysical." Between these two camps there is all the difference in the world.

The difference between the position I occupy and that of the emergentists with whom I cannot join hands lies in their respective relations to the doctrine of supervenience. Some emergentists are friendly to the doctrine, others are indifferent; all of them view supervenience as an important diagnostic tool. On my view, any adherence to supervenience—any adherence that makes the notion of "levels" and level boundaries appropriate—renders the emergentist package inconsistent, and its adherent the target of my book. I reject the universal principle that everything supervenes on the microphysical, although I am prepared to agree to the proposition that *certain* features of things at a large scale (for example, mass) supervene on other features at a smaller scale. Certain emergentists—moving closer to the reductivist position, rendering their position in consequence more palatable to the "mainstream"—display allegiance to a more general principle of supervenience in their articulation of "levels," which are meant to designate a divide between what is dependent and what is independent in the relevant metaphysical sense. I say that levels and level boundaries must go. And supervenience of everything upon the microphysical, as the doctrine that makes the hierarchy picture apropos, must go with it.

3. Supervenience Must Go

Hardly anyone in the field of metaphysics these days has a problem with the notion of supervenience as such. The term is at this point in time so well worn, no one has to explain it any more. But since it is in fact my target, I will do so. The idea is that you carve out two classes of traits or characters: the A class and the B class. The B ("higher level") traits will *supervene* on the A class whenever the A class is sufficient to guarantee the former: whenever you have a system with all the same A traits, you'll also be in possession of a system with all the same B traits. So you can't have a difference in the B traits without having a difference in the A traits as well. The B traits "emerge from" or "depend on" the A traits, because the latter "necessitate" the former. The terms "emerge," "depend on," "necessitate," and "supervene" now belong to a circle of mutually interdefinable ideas.

Still, in its usual articulation that we have rehearsed in the foregoing paragraph, the supervenience idea is not the conception that its most inveterate users mistakenly take it to be—it is *not* an asymmetrical notion such that when one says that A supervenes on B, one can be confident of also conveying the message that it is positively false that B supervenes on A. Still, its users are incorrigible:

they persist in using the language of supervenience as if it succeeded in being a language of asymmetrical dependence. But their error is easy to prove: the definition of supervenience guarantees simply that when A-characters are held fixed, B-characters will be fixed as well. But nothing in the definition prevents the reverse being true in a specific scenario: when B-characters are held fixed, A-characters are held fixed as well. Thus supervenience is *not* an asymmetrical relation. End of proof. This is decisive, in my estimation, in showing that supervenience language is inappropriate where we wish to discuss true dependence.

Marie Antoinette might have been gratified by this point: she recognized that not having bread did not imply not having cake. Why should it? Failure of possession of X—as she might say—does not usually come with implications about other failures of possession; why should failure of bread possession be any different? This *Marie Antoinette principle* therefore requires someone to explain why not having bread enjoys implications that not having X doesn't. However, dependence conceptions *must* be asymmetrical: every statement to the effect *X depends upon Y* comes with the implication that *Y does not depend upon X*. This is the point of articulating a theory of dependence. And the basis for the implication must itself be rooted in a theory of dependence *for that relation*. The advocates of supervenience owe us a great deal of work if they should like us to accept the supervenience notion in general philosophical usage as asymmetrical: at least they owe us qualifications of when supervenience claims enjoy the relevant implications—keeping in mind the Marie Antoinette principle. (Incidentally, I shall present a theory of dependence for the core dependence relations in science—which complies with the demand that dependence relations be asymmetrical. Look for it in chapter 7.) But never mind. We will simply overlook this failure for the nonce. Ultimately we will jettison the notion of supervenience as diagnostic of levels (for instance, as marking boundaries), a notion that the emergentist requires, on the basis that it cannot do the kind of conceptual work they are constantly calling on it to do. Still, no harm will be done by overlooking some of its less relevant faults at this time.

Now it is no small thing even in philosophical circles to say that everything supervenes on the microphysical. In particular, it is no small thing to say this—while also affirming novelty and downward causation. What I mean by that is that commitments are accumulating, and with them the possibility that consistency might be forfeited. In fact, that is the case here: the quartet of (1)–(4) is inconsistent. This is the basic complaint I shall be lodging: the picture is unintelligible. What remains of this section engages in a rather detailed way with literature on the topic of metaphysics on the subject of emergentism. Readers who are not interested in the minutiae on this matter, who wish to pass through the internecine dialog on this subject lightly, or who are prepared simply to take my word for it, may skip to the following section (page 47), where they can pick up the narrative thread where we leave it off here—having made the case that the quartet comprising principles (1)–(4) is inconsistent and therefore unintelligible.

Emergentism's most vocal critic for many years has been Jaegwon Kim. Kim often notes that emergentism markets itself as being an intermediate philosophical position between outright reduction, on one side, and full dualism (for example, Cartesian mind-body dualism) on the other. The stance is meant as a sort of compromise. This characterization of emergentism's position in the philosophical space also makes clear why levels are both prominent in the doctrine, but also thought to be neutralized—held harmless to physicalism. This fact accounts for its fortunes among philosophers and scientists—its fortunes are anticorrelated with those of reductionism—because reductionism is no longer popular. If Kim (1999) is to be believed, the preponderance of philosophers are advocates of nonreductive physicalism in the form of emergentism:

> The fading away of reductionism and the enthronement of nonreductive materialism as the new orthodoxy simply amount to the resurgence of emergentism—not all of its sometimes quaint and quirky ideas but its core ontological and methodological doctrines. . . . It is no undue exaggeration to say that we have been under the reign of emergentism since the early 1970s. (Kim 1999, 5)

Rather than diminishing in interest since Kim made that statement a dozen years ago in his stinging critique, emergence has only grown.

If emergentism were misconceived, in line with Kim's sharp criticism, then it might well be thought that the present book is superfluous, like shooting a horse that's drawing its final breath. And if emergentism is seriously viable, in spite of Kim's sharp critique and apparently having survived it, then it seems once again that my target is ill-chosen, since emergentism is so hardy.

I am of the opinion that Kim took aim at what he considered the strongest version, the one closest to reductionism—that version embracing all of (1)–(4)—and roundly defeated it. He is right: the version of emergentism at which he took aim is in fact not viable at all. But while Kim's critique contains all the ingredients for demonstrating the un-cotenability of (1)–(4), Kim did not press home his advantage (because he is much more conciliatory than I am, and perhaps also because of his own loyalties to the notion of supervenience as an important diagnostic—diagnostic of the existence of levels as well as where they lie), and thus did not light emergentists the way out of their error. In other words, he did not tell them that emergentism is really irreparable. He did not say that—as I maintain—emergence is reductionism in sheep's clothes. Kim, even after his scathing criticism, continued to encourage emergentists to articulate a more intelligible conception, one that nonetheless still retains the notion of levels and the all-important premise (3)—supervenience. This was his mistake. And so now, while I have come clean about those sentiments I share with certain emergentisms, I am nonetheless obliged to align myself with the letter of Kim's case vis-à-vis emergentism's viability, rather than with its conciliatory spirit.

I want to be very clear about my message: emergentism, as articulated by the conjunction of all of (1)–(4), is simply inconsistent, and therefore untenable, even unintelligible. And positively the only way out, while clinging to the sentiments regarding activity at every scale (a sentiment I heartily affirm and hope that at least some emergentists share with me), is to dispense with the notion of levels altogether—and so with the very core of the idea of emergentism. Showing that the notion of supervenience is not a useful diagnostic—in particular, it is not a diagnostic of levels, or characteristic of their relations—will occupy most of this chapter. In my stance—of dispensing with supervenience as a useful conceit, and furthermore of dispensing with levels even if we can get at them supervenience-free—I believe that I am, quite frankly, alone. But of course that's why I'm writing this book.

Kim's Critique

Rather than entitling his rightly famous essay "Making Sense of Emergentism," Kim might better have called it "Making Nonsense of Emergentism," because his arguments point out an important flaw in the emergentist package of (1)–(4) that brings that doctrine to grief. Kim's recommendations, pursuant to his arguments, differ from my prescriptions, primarily because of his commitments to levels boundaries. In the end, Kim challenges emergentists to come up with a better way of reconciling downward causation with a commitment to supervenience in a way that "unifies" a range of different cases.

First, Kim makes what he takes to be a simple observation that is meant to follow from the simple idea of supervenience itself: if some higher-level trait—call it T^H—causes some lower-level trait T_L, it must do so via causing the basal conditions of T_L. (This observation must hold, according to the principle of supervenience, whether T^H and T_L belong to the same system or to different systems. There may exist no common names of these basal conditions T_L in current vocabulary.) Of course since we are talking about causing basal conditions, these must in turn be caused by the basal conditions of T^H—let's call these $T^H{}_L$.

So then Kim (1999, 31ff) asks a simple question: why can't $T^H{}_L$, the basal conditions of T^H, displace the traits themselves as putative causes and/or effects? The idea is quite general: why can't the basal conditions of any traits or characters we wish to talk about displace the higher-level traits in any purported case of downward causation? Why can't the basal conditions do all the causal work? It seems that the answer must be that they can. This idea takes all the air out of the idea of downward causation. It leaves nothing whatever for higher-level traits to do, by way of causing anything. Thus this argument is proof enough that there is no room for downward causation, if you accept the principles at work in articulating the suggestion—primarily the principle of supervenience. Downward causation—the very pulse of emergentism-is superfluous and has no place in a "scientific" conception of the world. Presumably there is something pleasing about this result: the physical world is closed to causal action from above; hence the physical order is complete. The universe is single-scaled.

As I see it, this is an airtight case for the inconsistency of (1)–(4). Emergentism aspires to an antireductive stance, but it is still implicitly committed to a single-scale universe. Now it seems as though Kim himself is committed to (1) and (3), so it's clear to him which item is the culprit of the lot of (1)–(4). In his self-effacing way, Kim does not frame his conclusion in the form of a knockdown proof, but rather suggests that emergentists require a new sense of novelty, or at any rate an account of how higher levels emerge from lower levels in a general way, so as to give some plausibility to the program. I am not so sanguine. But if you, the reader, agree with Kim that causation (whatever it is) must be the glue of the universe, *and also* that causal powers must all lie with the basal conditions of things, no doubt you must see no hope for emergentism either.

Pushback

It might therefore come as a surprise to a reader innocent of the philosophical battles in metaphysics that there are quite intelligent people who retain their faith in emergentism. They argue roughly along these lines: once you understand that the basal conditions of any higher-level trait cannot possibly be all simultaneous, then you will see that either (1) the causal interactions among conditions or (2) the complexity of the whole causation business requires that supervenience be abandoned.[3] In other words, they give up (3) rather than (4). And they advance their stratagem as a virtue: the failure of supervenience in these crucial cases (where "novelty" is encountered) is the very signal of an emergent reality—a *new level*. "Novelty" becomes "failure of supervenience." While this maneuver removes the architects of this strategy from the crosshairs of the Kim criticism (and I would offer congratulations), it puts them back in the

[3] O'Connor (2000) rejects synchronic supervenience explicitly: "emergent states do not in general supervene on physical states" (2000, 112). S. Mitchell (in private correspondence, and less explicitly in her 2009) views a supervenience view as essentially reductionistic, and so withdraws from the language of supervenience. Instead she describes emergence this way: "What is required to make sense of scientific emergence is a higher conceptual framework that accommodates the dynamics by which the behavior of the component parts of a complex system generate a higher-level property as well as how that higher-level property in turn can causally influence the very components from which it has been generated" (2009, 44).

line of fire vis-à-vis the original challenge: in what sense are these "novel" traits really and truly *emergent* from the nonnovel? This challenge is really to say: why does the trait belong in some higher level? My answer—and I offer it here on behalf of would-be compatriots—is that there are no levels. But no doubt some of these emergentists will not like this view. (Those who do should simply switch sides and be done with it.)

But I would not go as far as some of my would-be compatriots do down the path to destruction. They have misplaced loyalties to certain conceptions about causality—as several chapters downwind of this one will detail at some length. I have no problems resisting the notion that a certain (as they have it, "higher-level") trait causes other traits by "causing" the basal conditions for them. I see this as akin to resisting the proposition that one can make lemonade only by arranging organic molecules together in a certain pattern—or, even more poignantly, by starting with quarks and electrons in a certain ratio, and following a complicated pattern of arrangements. One does not "cause basal conditions," rather one squeezes lemons and adds sugar and water. The former activity (if one could do it at all) would be a cobbling of lemonade; the latter is a resheathing of stuffs on a (very slightly) smaller scale to make stuff at a larger scale. That will be the subject of the next chapter.

More philosophically, the glue of the world is not causation—there indeed is no single all-purpose glue. There are many kinds of glue—and this too will be a recurring theme of the present book. (Importantly too, the name of "causation" should not be affixed to glue that holds things together; that glue is "dependence.") But more important still, the "glue" of the universe is not fundamentally vested in the lowest level, even if we could be certain there is just one kind of glue and there is a lowest level. This too is a theme to which we will return often. Some of my would-be compatriots are too invested in the notion of supervenience to relinquish their false conceptions regarding causation. But Physics itself testifies against this notion that activity occurs only at the micro level—as most of the remainder of this chapter will discuss. I have in mind both quantum and classical physics—and in the latter category classical physics in the tradition of analytical mechanics.

Of course not everybody will agree with me about how to understand quantum physics or analytical mechanics. But some of what I will be saying—such as for instance when I point out very general failures of supervenience (in the next section)—there can be no arguing with. Still, someone could rigidly commit to the principle of supervenience, insisting that we simply haven't identified the true "basal conditions" from which the higher level somehow emerges. This is the "more research is required" gambit.

A different stratagem is to hold on to the idea that a complete lower-level description will always be an incomplete *description* (perhaps even a distortion) of the true entity—that, in other words, there can be no complete description of the higher level in terms of the lower level, and that insisting on the completeness of the lower-level description distorts our conception of the science that is performed when higher-level (truly higher level) properties are properly introduced.[4] But this is simply an *insistence* on emergence, not an explication of it. Furthermore, it does not say why there can be no complete description at the lower level—if there truly are levels.

One could, in trying to support this idea through giving some explication of emergence, say that the emergence of upper-level properties from lower-level properties is a matter of causation of the former by the latter, not a matter of composition or realization, after the fashion of Humphreys (1997b). So, for instance, one of my favorite examples of an emergent character (and I consider it nonsupervenient as well): the rigidity of a piece of wood or steel bar, one might say, is "caused" by the properties of the molecular components—it is not simply a metaphysical sum of them. Rigidity is an activity that is produced by rods because of how the materials of which they are composed came to be, though each material, whether metal or wood or what have you, requires a different story.[5] This position might cause some discomforts of its own. (How can we discover such causes? Can causation between lower and higher properties differ across instances—same causes but different effects on different occasion? etc.). Still, it's not untenable, and it does *squirm* out of the difficulty posed by Kim. But—and here's the rub—it gives up entirely on supervenience, making of causation absolutely everything. And everything I've said thus far might suggest I should be amenable to such a view. I am—up to a point. I don't hold with using the notion of supervenience as a diagnostic. Philosophers far and wide are wedded to it for the service it renders in making us all physicalists, but what's in a name if the doctrine has no substance?

Adherence to supervenience seems to have the quality of a pugnacious challenge: "just find me some information at a higher-level H that is not contained in the description of the lower-level L—just one." You'd think that would have been done if it were ever going to be. And yet it seems not to have been. So it seems as though adherents to the supervenience doctrine have the upper hand. But the logic isn't as simple as the pugnacious stance suggests. There are examples where the pugnacious challenge meets its match, but there seems always to be a way out. Here's what I mean.

Someone—let's just say it's me—suggests a piece of information: I suggest rigidity as a candidate piece of information that fails to supervene on information at more micro scales. The challenger then pushes back: yes, we can retrieve

[4] S. Mitchell (2009) occupies this position, but nonetheless affirms our ability to make sense of the higher level in terms of the lower level: "What is significant for understanding the complexity of this system is to see how an adaptive structure at the higher level (nectar supply) emerges from individual behaviors and how the higher-level structure causally influences in a feedback loop behavior at the lower level" (43).

[5] This claim is suggested, though not explicitly articulated, by S. Mitchell 2009.

rigidity information by querying the "molecules" of the rigid substance about the intermolecular forces and such. "OK, go ahead," I say. They say: "we can." "All right, proceed," I say. Nothing is forthcoming but a staunch reiteration of the party line. That's all they can do. There is nowhere in the physics literature an account of rigidity solely in terms of actual properties of molecules.[6] What my opponent really means is: "Show that I can't." But of course this is a very different sort of challenge. For this means proving that there is no theoretical function that could be devised, taking in all the many magnitudes available to my opponent at that micro level, that would (under all the relevant conditions) output a figure that would be corroborated by all known ways of making measurements of rigidity. This is a monumental, if not outright impossible, feat to perform. At least betwixt breakfast and supper. And any claim to having accomplished the feat is subject to the retort that I simply haven't examined all the possible functions, and indeed that scientists may not yet know all the facts about the constituents of things. So we arrive at something of a stalemate on the topic, in spite of the appearance to the contrary.

In the end no position on supervenience—whether mine or that of my opponent—enjoys an upper hand. There are clear problems with the idea that any position on the subject of supervenience deserved a "default" status. This is why I will pursue a different strategy in the remainder of the chapter. I will focus on a single example that I fervently hope will demonstrate why the supervenience test cannot diagnose what is wanted.

Before we embark on that, it is worthwhile just reminding the reader that supervenience is by no means the best way of being a physicalist. There is another, easier way of adhering to the stouthearted materialism of our Vienna Circle predecessors. This is just to declare everything physical and raze the so-called levels. That is the strategy I should like to promote. In promoting it I'm following the evidence. Evidence, as I shall argue, points away from supervenience as the relationship between scales of measurement—away from the sedimentary model of macro traits—and toward a quite different way of thinking about how things hold together. I propose to say that some things are *integrated*—that they are *integrates*; while other things are merely *aggregates*. That idea will make its appearance in due course.

4. Supervenience Fails, but This Is Not Diagnostic of Levels

Some of the most spectacular failures of certain traits to supervene on certain other traits reside (rather perversely, as it might seem) in microphysics. I wish to display such an instance. This is the first instance on display in this book of a

[6] I discuss this example at some length in Thalos (1999a).

scale-free universe—of activity happening at a higher scale of organization or aggregation. But it's not as though no one has noticed the particular case I will discuss: the puzzling case and others like it have been common currency for many decades. And the failures of supervenience (known by a different name, as will become clear) ultimately provoked invention of a whole branch of physics to handle them—namely quantum theory (or quantum mechanics, as I'll sometimes say). It has been the most astonishingly successful theory of all time, welding together into the same entities wave characteristics with particle characteristics. Philosophical recalcitrance vis-à-vis failures of supervenience, while in some ways understandable (because nature has brought our minds up to be classical), is nonetheless quite puzzling. But let's not get ahead of the story.

Let's get clear first, though, about the point of the story. Its purpose is to display an astonishing failure of supervenience and to show that this failure demands new theorizing about the physical. This case will also make it clear that a failure of supervenience is not in any way diagnostic of the levels spoken of in emergentism. In other words, the failure of supervenience cannot be a test of where boundaries between levels lie—it can't provide us any kind of Geiger counter for levels. For instance, if you held to the idea that novel higher-level traits supervene, but not on instantaneous values of lower-level traits, you might have thought it attractive to use supervenience as the test of novelty.

So the point of this exercise will be to demonstrate that supervenience fails where you least expect it, and this cannot be used as a way of marking levels either. You'll recall, of course, that some aspirants to emergentism deny failures of supervenience between levels—just as Kim advised (as he himself leans toward reduction). So their response to failures of supervenience will obviously be one of simple puzzlement, whereas the emergentist who positively embraces failures of supervenience (as a diagnostic of levels) might be prepared to adopt these as evidence for levels. The case I wish to display will make both emergentists' stances unattractive.

The failure of supervenience in this case is also simply astonishing. Why? Because the characters that will be revealed as nonsupervenient are not even obvious candidates from anyone's perspective for being "novel." In other words, if one were to be anticipating proposed counterexamples to supervenience, one would never have guessed that we'd find it here. The case I'll be displaying is not an example *anyone* would have chosen of novelty. And yet, there it will be: a property of a pair that fails to supervene on the properties of its constituents. This is one of the reasons that we can't use failure of supervenience to identify the positioning of levels. Since then we'll have no diagnostic whatever, levels become extremely "theoretical" or "philosophical" in the most derogatory sense of the word.

The portent of this case is tremendous: it is as clear a sign as we can have that we need to get off the emergentist's sinking ship *here,* where it's absolutely crystal clear that facts at one scale of things do not guarantee anything about facts at a slightly larger scale—an unambiguous demonstration of the fact that we cannot count on any kind of reduction between any two scales of measurement—before we are thoroughly misled as to how parts and wholes interact to make larger unities.

I'll be considering an instance of what often is referred to as *entanglement* or *EPR-type correlation* (so named for the authors who in 1935 first conceived of the setup in the form of a thought experiment: Albert Einstein, Boris Podolsky, and Nathan Rosen). This can be easily grasped, so no panicking just yet. To keep things very simple, let's discuss a very basic example of a pair of particles—their identity is unimportant at this time. Still, handling such cases appropriately requires facets of quantum theory. I'll introduce the case by introducing experiments in which the puzzling behavior is manifest. Two spatially separated detectors in our large lab have been set up to measure, in a variety of directions, the "spins" of particles approaching them in real time. Figure 1.1 depicts the setup in our lab. There are two wings: the A wing and the B wing. At each wing a detector can be set to measure either spin in the horizontal direction (the direction of X_A in the A wing, or the direction X_B in the B wing) or in the vertical direction (Y_A or Y_B). For instance, measuring the spin of the particle in the A wing in the vertical (X_A) direction will lead to one of two possible results: either the particle has spin Up or it has spin Down. No other options are possible because the characteristic is quantized—which is to say that it can only take on one or the other feature (just as an eddy can swirl clockwise or counterclockwise and there are no other ways of swirling). Up and Down just refer to the way we designate spin

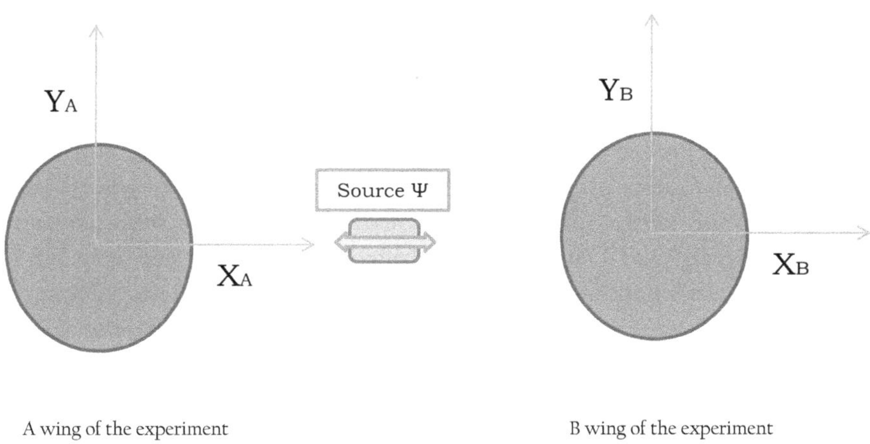

Figure 1.1 The EPR-type experiment in our lab

"direction"—*Up* designates the direction of the X_A or X_B or Y_A or Y_B arrows as shown in the figure, while *Down* designates the direction opposite.

When we leave the setting unchanged (on either detector, in any direction we choose), we get Up half the time and Down the other half. The particles thus seem to have no "bias." We are then told that all the particles we've been catching have been prepared together at a certain production site, that they are emitted in pairs, one proceeding to the A wing of our setup and the other proceeding to the B wing, and that care has been taken to ensure that each emitted pair differs from the other in absolutely no way (and this latter is confirmed by a wide range of experiments on the arriving particles). We are therefore justified in supposing that each measurement being taken (when we don't change settings) is of the same thing. Taking these two facts together, it accordingly makes sense—and it is in fact quite consistent with other principles of quantum mechanics—to proceed with assigning equal probabilities (of $\frac{1}{2}$) to the particles reaching each site for displaying Up and Down spin characteristics upon being measured. As routinely said, there is nothing more to be said about the individual particles in respect to their spin features.

And yet there is. While it is quite correct to assign the aforementioned probabilities as we discussed, and consistent with every possible experiment we might wish to perform downwind of the measurement at each wing of the experiment, it is also appropriate to inquire whether there is relation between Up and Down findings at the *different wings* of the experiment. And the results of this inquiry are enormously revealing.

Suppose we are able to keep track of which particle arriving at the A wing was initially paired with which particle arriving at the B wing. Obviously we can keep track of which direction we're measuring spins against (say whether in the vertical or horizontal direction, of the plane at right angles to the line of particle motion). Given what we have said about the particles, what probabilities does it make sense to assign to the joint event of an Up measurement in the A wing being paired with an Up measurement in the B wing, when both detectors are measuring in the same direction (say vertical)? We've said that there is nothing more to be said about spin in that direction except that it's going to register Up at A (or B) with probability $\frac{1}{2}$, and similarly that it's going to register Down at A (or B) with that same probability. So the (joint) probability of getting two Up spins in the vertical direction, one at each wing, must be some nonzero number between 0 and $\frac{1}{2}$. It can't be more often than Up at just one wing, but it also can't be zero, because surely Up at A and Up at B must sometimes co-occur, if they can occur at either wing, and if that's all we know about the occurrence of Up.

To see why, consider tossing two identical fair coins simultaneously. Each one has a probability $\frac{1}{2}$ of coming up heads and probability $\frac{1}{2}$ of coming up tails on any given toss. Now toss them together. What's the probability *both* will be come up heads? The right answer for fair coins is one-quarter—the product of $\frac{1}{2}$ and $\frac{1}{2}$.

As we say when we make the calculation, the tosses are "independent" (which is itself something that is surprisingly hard to define, and with good reason—but that's a topic for another book).

Entangled paired particles, no matter far-flung, are nothing at all like coins, however. The fact is, we *never* get such joint events as Up/Up. The particle spins in each direction are *perfectly* anticorrelated. And more importantly, we can't get that information from knowing what's happening (and what could possibly happen) locally at each of the wings. In some sense this really isn't surprising. That is because to make the prediction about coins, we had to know whether they were "independent." If your coins were always anticorrelated, you would suspect that their tosses were in fact not independent. But then how could you make the prediction that each trial on each coin is as likely to turn up heads as tails? To be able to combine the two predictions (perfectly anticorrelated, but also as likely to turn up heads as tails), you require a quantum-theoretic conception of things!

That was just a bit too quick. It's true that combining the two predictions requires a quantum-mechanical framework. To see why, consider the reasons why we can't use a classical model. Consider a classical example of a perfectly anticorrelated pair of objects: a pair of gloves. (In fact this was a favorite model of Albert Einstein's.) From any pair of gloves, drawing a Left glove always leaves behind the Right, and vice versa. And there's nothing at all mysterious about it: it's because gloves stand in this relation to one another that they are anticorrelated in their pairs. Thus one might wish to conceive of paired particles with entangled spin on the model of gloves.

This model, however, works only up to a point. Suppose we think of the paired particles on the model of gloves. The pair-generating device throws out a Left and Right randomly toward the pair of detectors. It's as likely to throw the Left toward the A wing as toward the B wing; but whichever wing toward which it throws out the Left, it must always throw out the Right toward the other one. The model is capable of generating the anticorrelation. And now to get the other prediction: each trial at each wing is as likely to turn up Left as Right. Obviously this can happen if we simply insist on the generating device randomly distributing Left and Right between the two wings. That's where the glove model will succeed. But further elaborating the model along the lines we have used (the gloves) so far, so as to model the other characters in the system, fails miserably.

We need to elaborate on the glove model to render it useful in situations with more than one character (with values in surplus of Left or Right). When we run experiments with real paired electrons in the lab setup we've currently got, we can measure any one of two characters, because at each wing we can measure spin in either X or Y direction (actually, we could measure spin in each of the three orthogonal spatial directions). Since the present glove model gives us only one character, which can be either Left or Right, it requires enhancing if it will be up to the job. So consider that gloves can also be either White or Black: whenever one is White, the other must be Black (W/B). They can also be Leather or Suede (L/S). So any given glove might be found to have a handedness or a color or a fabric—any one of three characters. Since we can measure different characters at different wings of the experiment, the statistics will of course be more involved. To get the predictions right, not only must the enhanced glove model generate half-Left/half-Right, but it must also generate half-Black/half-White and half-Leather/half-Suede. Of course there is no problem with insisting it be able to do so. This is the rub, however: when the glove model does what it is so well suited to do—when it delivers on the anticorrelations between any given character (handedness, color, and fabric), it will *always* mispredict cross-character statistics. For instance, it will mispredict the proportion of Black-Suede, and White-Suede, and Left-Black. It is a theorem—indeed, it is the substance of the famous Bell's Theorem (Bell 1964)—that a glove model of the sort we have envisioned will always mispredict the cross-character statistics (color/handedness, for instance) of entangled particles, whenever it manages to predict correctly the statistics of the same character (color, for instance) at different wings of the experiment.[7] And here's why: a glove model (by design) always assigns a determinate color and handedness and fabric to each glove—that, after all is the logic of a classical model. But this logic always constrains the statistics. The plain fact is that the constraints are all wrong when applied to entangled particles—that is what experiments verify. Experimentation verifies quantum predictions to mind-boggling precision and disconfirms classical models to that same precision.[8]

We need a model of entangled particles that is capable of building in the right correlations—the right statistics—between the different characters. It can't be constrained to the sort of one-size-fits-all dependences that classical theories always build in. Classical models have a certain logic for building dependences. But that logic is wrong. To capture correctly the sort of correlations that we find in entangled particles, a model has to be able to build in correlations between characters more flexibly than classical models can do. In quantum theory, the description of a system (what we call in quantum mechanics the *state function*) is designed specifically to contain all there is to know about a system.[9] But the information it can contain can be structured considerably more flexibly. To illustrate this point, notice that the glove model can create correlations (global

[7] I won't enter into the intricacies of where precisely the model goes wrong; they are not terribly illuminating. But I invite the reader to consult Cushing and McMullin (1989), Mermin (1981), and Albert (1992).

[8] See Mermin (1981) and Albert (1992) for further discussion of the statistics.

[9] In classical models, a character is a much less formal thing; a classical model is an analog model, whereas a quantum model is *digital*.

characteristics) by assigning determinate characteristics to each member of the pair (local characteristics); the correlation then flows from (supervenes upon) the local traits. Quantum models do not require that correlation information be constructed in this way. Quantum models allow the correlation to "float free" of the values of the correlated characters. Still, we can "query" the quantum state function for information about any feature we like—local traits *and* global ones. When it comes to paired particles, quantum mechanics has just *one* description for them—it describes them as a linked pair—as dependent in a specific way. So you will get information about *joint* probabilities—from which you can infer information about each of the events separately (by a kind of courtesy of logic).

When we query the "pair" quantum state function, we can ask: what percentage of the time should we be seeing correlations/anticorrelations? And the answer is straightforward: the particle spins, in any direction you please, are always anticorrelated—*with probability 1*. This is information we can only ask when we ask about the "joint" probabilities (which of course makes sense; how else can you get such information?). So it seems that at the scale of being a pair, however far-flung they might be, the particles have a further feature of their spins that isn't information contained in the "reduced" or "local" features (or the way we talk about and measure local features) of each member of the pair. There is more information about the spins of particles in the state function of the pair than is contained in the information we can give about the (separate) features of the individual members of the pair. Indeed, quantum mechanics does *not* assign to the individual particles separate wave functions. Their state is a state that applies to them *strictly* as a pair—but even so the wave function, by courtesy of logic, has a good deal to say about what we will find when we treat the parts as individuals, measuring them accordingly. And it always gets everything right.

Of course if quantum mechanics didn't get everything right, we would have no philosophical problems. We'd be able to shrug it off as a kind of fluke. We have problems in devising analog (classical) models of quantum behavior because the formal machinery of quantum mechanics does so a lot better—so much better than any other calculus we can use to make predictions in these contexts. And, furthermore, because it seems that when we "query" quantum mechanics' characterization of the state of the pair, it tells us things we would not have said about it, commonsensically. One of these things is the failure of supervenience of probability distributions for joint events upon probability distributions for individual events.[10]

[10] Another of these things is the failure of a "common cause" as an explanation of the joint distribution. If your coins were always anticorrelated, you'd suspect that the outcome on the one caused the outcome on the other; in cases of entanglement, this solution is ruled out by the principles of relativity because we can specify that the outcomes are measured at precisely the same time—and the results would be identical.

According to quantum proprieties, one derives the probability distributions for the local events from the joint distributions for "entangled systems," and *not* the other way around: the former apparently depend on the latter, and not the reverse. So if supervenience is appropriate (which is highly questionable), the relationship is the reverse of what emergentists (and card-carrying physicalists generally) would suppose. This is astonishing.

Take a moment and let it sink in just how astonishing this is. There is a calculus that, when queried, can give us complete information about the probabilities (joint probabilities) of certain events co-occurring. And this very same calculus can also provide us probabilities about the same events when taken individually. And when we query for both sets of information, we can work out from the result whether or not the events (and by implication the associated entities) are dependent. It only stands to reason, given this characterization of things, that the information about the joint probabilities is the more complete. The local information has to be less rich *by necessity*. In some sense, we should have been open to failures of supervenience. But somehow, we never were. We simply *assumed* that if entities are flung far enough apart, everything one needs to know about them will be local. (This is the principle of locality which is discussed in the next three chapters.) But that just is not the case. Locality is simply not universal. And in itself this is a good reason for being suspicious of causation as the cement of the universe—which will be a major theme of chapter 4.

To reiterate: there are systematic failures of supervenience of aggregate properties vis-à-vis properties of constituents, so that all the information we can ever collect about the constituents will always be insufficient to generate a correct prediction of the aggregate feature. Or at any rate, a quantum theory is poised for any such eventuality, whereas a classical theory cannot be. When such an eventuality is actually occasioned, we can only theoretically predict the former from the latter—predict the individual properties from the aggregate property—and not the other way around. This would suggest that the aggregate property is the independent character—precisely counter to classical common sense![11]

[11] Some readers of this book might be thinking that there is a certain move open to the champions of supervenience as defined in the text, that nothing said so far contravenes. This is the move according to which if we simply add *relations too* into the supervenience base, we can preserve the principle. For instance, if we were to add handedness relations, or whatever other relations obtain, for entangled particles, in their cross-characters, we would essentially be able to provide the same information, but at the supervenience base. Of course if we are willing to admit into the supervenience base relations purely, without need of the categorical properties (in classical theories) that ground the relations, then that of course would save the doctrine. But the new doctrine would be unrecognizable to its original patrons. Indeed, the new doctrine looks considerably more like the scale-free model that I am defending in this book—and no doubt will encounter considerable opposition from those who would like to preserve the doctrine of supervenience.

What's fundamentally important to note here is this: all the information that quantum mechanics gives us about the system is coming from one and only one source—namely, the state function of the *pair* of particles; in quantum mechanics there is no *other* source of information about those particles whatsoever. We can query that state function about anything we like, and it will give us straightforward answers. What will we find upon measuring the spin, in the horizontal direction, of the particle in the A wing? "Up (probability $\frac{1}{2}$); Down (probability $\frac{1}{2}$)." And what will we find on measuring the spin in the horizontal direction of particle in the B wing, upon finding its counterpart in the A wing to have a horizontal spin of Up? "Down (probability 1)." We get straightforward answers to these queries. In fact, the state function is (at least theoretically, if we've set up the problem right, and solved it correctly) the only source there could possibly be for information about the particles we subsequently query in the different wings of our lab. If there's something inconsistent about the information at these different levels of generality, it would prove absolutely disastrous for the theory. Indeed there are many theoretical demonstrations to the effect that no state function can possibly be inconsistent, once the elements of the theory are properly interpreted.

By way of preliminary commentary, notice that not once did the issue of "novelty" appear as relevant in this case. It's not even clear we're getting different or additional properties at the scale of the aggregate (the pair). True: the property of "correlated" or "anticorrelated" is in some sense new—but not in the sense emergentists hope for, because this property should be construed as either "logical" (Up here *and* Down there) or simply structural (if you think of the pair in spatial terms). In quantum mechanics, we can manufacture new symbols (names) of aggregate properties by mathematically composing the names of the constituents. And only these can query the state function of the aggregate. (We query the state function of the pair for "local" characteristics, by composing the name of the local variable—let's say it's L—with a kind of "dummy" for the other, for example: $L \otimes I$.) And the state function that quantum theory associates with the particles flying about our lab says that the particles are perfectly anticorrelated. We understand that result perfectly: spins at opposite wings are opposite to one another—in the representational scheme we've devised, they must always sum to zero. But why? That's a question a scientist should of course wish to answer.

And the answer is: it's the result of a conservation principle. Conservation principles are *global* rather than *local* matters: they have an "everywhere" or "total" character about them, not restricted to acting "close by." That's the distinguishing feature of failures of supervenience: where we find them, some kind of principle of global organization is not too far behind. Failures of supervenience are often accompanied by failures of the principles of locality—for example, of failures of our capacity to pin the anticorrelation of spin on the fact that one particle influenced or acted on the other via a "message" or action exchanged between them.

So we can have profound failures of supervenience *even when no novel property is on the scene*—nothing in any conceivable way new from the perspective of making measurements. This is important now for one reason. We've already noted that the emergentists' (1)–(4) are not jointly tenable. A certain cadre of emergentists might, at that point, have rejected supervenience as a universal principle and used that failure to good effect. They could have marked their level by the telltale signs of "novelty" (a failure of supervenience) and simply joined the ranks of those who acknowledge failures of supervenience among levels. But there is no real traction here for this lot of emergentists, however much one may admire their courage: failures of supervenience are not always accompanied by novelty.

What does the failure of supervenience here portend? The truly important lesson is that locality is not a universal feature of our universe. The universe doesn't seem to have to obey familiar locality principles. (It obeys local conservation of *probability*, not actuality. This is a feature of the Schrödinger equation, and we will discuss this topic presently.) It's quite possible that when it does in fact, just by accident, obey locality principles, supervenience holds where we think it ought. But the universe can depart from obedience to local laws. And perhaps one day we will have an explanation of when it does and when it doesn't. We don't have an explanation now. Not even the slightest clue. And we can't appreciate that we don't, so long as we go on ignoring such failures of locality as we do have evidence for.

So the moral of this chapter thus far is: supervenience, when it obtains, is simply contingent (because locality, when it holds, is similarly contingent). Failure of supervenience is just a straightforward option that is open to the cosmos. The state function of the universe can be queried: are your parts coordinated or not? That is just a question that the cosmos can answer *coordinated*, without falling afoul of any physical laws—and while obeying other, higher laws.

5. Sync and Its Shrouding

What do laws or principles involving coordination look like, in general terms? I will call theories that allow for coordination principles *sync models*. They form a large class of models of phenomena that, basically, involve correlations (synchronies) across vast stretches of space and time—and in a way that strikes us at our scale as having been "preprogrammed" or else "worked out by like-thinking minds," neither of which can possibly be true of the entities we are talking about. What sync models do is formalize or code for the correlations in a way that obviates the suggestion that either real-time "preprogramming" or "intelligence" is required to achieve sync. What sync models illuminate is the fact that what typically requires sync, mathematically speaking, is suspension of locality. Mathematically, there's nothing all to the suspension of locality. But what does it mean from a metaphysical point of view? Much more than common sense is prepared to accept, it seems, if the resistance to sync especially among philosophers is a sign of anything.

"By tradition, physics is a strongly reductionist science," writes Paul C. Davies (2006, 35) while strongly adhering to that reductionism himself—to that "nothing-buttery" that once upon a time might have been styled, with swagger: There's nuttin' but atoms and the void. A physicist and physics popularizer, Davies has ventured with both feet (not always fully aware of how poorly shod) into areas inhabited primarily by the philosophically trained. Among his many exploits is a venture into the area of emergentism. Sometimes his attempts to inform and instruct on the reasoning of physicists makes these good folk sound like they're fresh from the nineteenth century—and a monolithic breed to boot:

> For the *physicist* [my emphasis], the only causes that matter are, to paraphrase Thomas Jefferson, the ones that kick. Wishy-washy talk of global cooperation is no substitute for observing a real, honest-to-goodness force that moves matter at a specific place. And if the movement is due to just the good old forces we already know about, simply re-packaged for convenience of discussion, the response is likely to be a monumental "so what?" (2006, 39)

In this passage Davies is talking about the physicist's conception of force and intending it as a kind of corrective to the philosophical notion of downward causation: there is no such thing as a new force that acts (as it were) "from the top down." But as we've seen, no one on the planet, not even on the emergentist continent, has this sort of proposal in mind at all. In fact, no one is talking about out-and-out *forces*. Still, Davies's point is instructive: physicists read "cause" simply as "force." This point will be important for us in the next chapter.

Davies's tone is unmistakable: physicists don't make the silly mistakes philosophers make. Physicists are a breed that know better. They are better anchored in their thinking. Their touchstone is locality: the holy grail for them is a theory of behaviors that is founded in the notion that causes (forces) act locally (close to home; so close in fact that they never act "at a distance," however short). Physical theories always strive for this, according to Davies.[12] And physicists are all of one

[12] And it rather seems to be the consensus that they have now stopped looking—since now we have quantum mechanics and it explains so much in connection with matter (even condensed matter): "With the advent of both quantum mechanical accounts of chemical bonding and the explanations of biological phenomena of inheritance by the biochemistry of DNA . . ., the properties of chemical compounds, like water, and many features of living organisms have arguably been explained in terms of the properties of their constituent parts" (S. Mitchell 2009, 25). "The emergentists were wrong in thinking that sundry chemical and biological properties were emergent; but this was an understandable mistake

mind on the matter. That is how the passage reads. What Davies does not tell the reader is that physical theories do not always adhere to this principle, however much physicists might strive for it. Why doesn't he say this? It's really quite puzzling, because just a few pages earlier Davies notes that physicists are really not all of one mind, not even concerning the "nothing-buttery":

> Physicists who espouse anti-reductionism usually work in fields like condensed matter physics, where reduction often fails even as a methodology. These workers are impressed by the powerful organizational abilities of complex multi-component systems acting collectively, which sometimes lead to novel and surprising forms of behavior. . . . [T]he molecules organize themselves into an elaborate and orderly pattern of flow, which may extend over macroscopic dimensions, even though individual molecules merely push and pull on their near neighbors. . . . This carries the hint that there is a sort of global choreographer, an emergent demon, marshalling the molecules into a coherent, cooperative dance, the better to fulfill the global project of convective flow. Naturally that is absurd. The onset of convection certainly represents novel emergent behavior, but the normal inter-molecular forces are not in competition with, or even over-ridden by, novel global forces. The global system "harnesses" the local forces, but at no stage is there a need for an *extra type of force* to act on an individual molecule to make it comply with a "convective master plan. (2006, 36–38)

There are notes of derision in this passage—a suggestion that the "condensed-matter physicists" are not as stouthearted as they ought to be. (But read the passage again: Davies himself characterizes these systems as acting at a high scale by "harnessing," though there are no new forces. Does he mean this metaphorically?) Still the takeaway is that intimations from "condensed-matter physicists" are absurd. Because "the physicist" is bound by principles of locality.

And yet it isn't so. Quantum mechanics isn't so bound. The single most spectacularly successful scientific theory of all time isn't so bound. Given its stature in the physics pantheon, I nominate quantum mechanics as the premier model of *sync*.

True: quantum mechanics does not countenance "new" forces, ones not known before. And as I mentioned just in passing earlier, there is a kind of locality: there is local conservation of probability flux (just as there is local conservation of

given the state of the sciences before the advent of solid-state physics and molecular biology" (Kim 1999, 9). The problem is this: quantum mechanics does *not* explain the properties of things in terms of just their parts. That's the point I'm making here. And it seems that everyone on the spectrum vis-à-vis emergentism (from sympathizers to critics) has missed that point.

water through any unit volume in a river, even though water might be flowing quite turbulently and speedily through that volume). Neither fact prohibits the failure of locality in the more prosaic sense—the failure of global facts to be mere "sums" of local facts. Nothing (and certainly nothing Davies cites as the fundament of stouthearted physics) prevents failures of locality—or, correspondingly, failures of supervenience.

The quantum theory is that most vigorous of all physical theories, which so many people find unintelligible because intelligibility in their view demands locality in the sense of supervenience of global upon local. As a consequence, quantum mechanics violates fundamental axioms of causality too (for instance, the so-called *principle of common cause*)—a fact that itself has been proven beyond disputing—however much the stouthearted physicists of Davies's acquaintance still demand it. This fact *all by itself* guarantees that it will violate supervenience. This fact evokes some resistance. Indeed it prompts a kind of subterfuge on the other side.

Perhaps "subterfuge" is too strong. But suppose you were a "condensed-matter physicist" fully aware of the derision that is aimed by the ilk of Paul Davies at your own sort. You know that if you use the inflammatory language of "emergentism" (I would urge you to use *scale freedom* instead, by the way) your work will be subject to even harsher scrutiny above and beyond the harsh-enough scrutiny to which any new idea is normally subject. (Just take note that even in 2006 Davies can say "absurd.") Who needs that? So you shroud your theories in language you believe is innocent enough to pass. I've argued elsewhere (2006) that this was a practice among condensed-matter scientists for some time. Wilson and Kogut's beautiful 1970s theory of renormalization, dealing elegantly with the universality of materials in the critical regime, had therefore to be packaged as a reductionist theory—as all serious physics had to be (and indeed still does). I will discuss renormalization and extract principles for metaphysics from it in chapter 8. Just by way of enticing the reader to read on and not to skip that chapter, consider how Binney and coauthors write, in their important text on phase transitions: "there is a paradox here [!]: on the one hand it is clear that inter-atomic forces are responsible for the very existence of a phase transition; on the other hand the details of these forces cannot play any role in determining the critical exponents, since these stay the same when the atoms, and therefore the forces, change" (Binney et al. 1992, 21). They are shrouding their position in the language of "paradox." Is there a paradox? Or is it outright inconsistency to hold that both sides of the quoted sentence are true? Since it takes a philosopher to work out the difference, it is safe enough to use the language of "paradox"; it is the means to respectability.

For the time being, we will leave failures of supervenience as a spectre for those who insist that the universe is active on a single scale only—the micro scale, of course. We will conclude with what is more important about sync models—the positive element that they share in common: they involve a kind of "leakage" of information from point A to point B, even when these points are remote from one another; and the models are quite typically agnostic as to how the information is leaked. Sometimes "lags" in leakage are part of the model; sometimes the leakage occurs simultaneously. One might even say that "leakage of information" is the hallmark of quantum mechanics. "Leakage" is shared by most (if not all) of the examples discussed by one of the foremost scientists of synchrony in our time—the applied mathematician Steven Strogatz—in honor of whose gem of a book *Sync* (2003) I have chosen the label I am using for these models throughout. I will discuss many of his examples too, always with an eye toward the philosophical lessons. But first, some more philosophical squirming.

6. Sync Is Controversial

Suppose that you are still hostile to the notion of a scale-free universe, or simply that you can't accept, willy-nilly, failures of supervenience. You might even be an emergentist who nonetheless clings to your belief that all activity transpires at the minutest scales. What are your ways out vis-à-vis my example? I will name here three ways out (all the ways I can think of) and show you why they are bad choices.

1. Not a Level Marker

You might wish to insist that the difference between one level of aggregation and another, in our two-particle case (or even in those cases you suspect are waiting "in the wings" involving a huge number of entangled particles—such as for instance in superconductors or superfluids), cannot possibly amount to a "level" in the more robust sense that emergentists have in mind. We don't have the appropriate micro/macro boundary in mind. So the failure of supervenience here exposed is inconsequential. We are still only talking about the "physical" level. So there's no problem.

Let's examine the position more closely. You wish to insinuate that failures of supervenience, for instance those "below" your preferred line of demarcation, are of no consequence. We have already reviewed the difficulties of drawing any convincing lines of demarcation. But let that point go. Let's suppose that the preferred line of demarcation is considerably "higher" than the present line where supervenience fails. Why isn't the present failure problematic? You are now insisting that it's because everything we're talking about is *still physical.* I reply that mountains and planets are physical—indeed that there is physical at every scale of the universe. You shrug. Failures of supervenience, as such, don't perturb you. They would only perturb you if they happen to fall at your preferred line of demarcation. Only then would your preferred sort of physicalism be threatened.

This position thus collapses to a boundary-protection problem. It's not at all clear why you are not disturbed by inconvenient failures of supervenience—so long as they're conveniently located inconveniences. Your notion of newness, which perhaps you had originally thought might have been diagnosable by failures of supervenience, has now to be articulated independently of any notion of supervenience. It's not at all clear that your argument can be attacked—not clear what it is claiming so that one could attack it. And it is not at all clear that the notion of supervenience can serve you in any way in your defense of this position. This bulletproof quality to your position is the philosophical analog of the unfalsifiability of a scientific hypothesis. It makes your levels "theoretical" in the most derogatory sense. Not good news for you.

2. Weirdness Is Micro

In a more conciliatory vein, you might withdraw to the position that weirdness is a *sign* of the micro. We know that everything has to look "classical" in the "classical limit," after all. So the line you are drawing is not so arbitrary. It is the "classical limit." If I could show that there are failures of supervenience above the classical limit, you would consider your position compromised.

OK, let me take up the challenge. It's not clear where the "classical limit" line in "classical limit arguments" is supposed to run. But I'll let you draw it where you like. For my part now, I draw attention to entanglements at scales higher than your line. There are documented, numerous, very important quantum effects we can observe at macro scales. (And we'd so love to be able to take advantage of more of these in development of technology.) Lasers and superconductors are two important cases. We'll turn to these macro effects in the next section. But these cases demonstrate the deep instability of your "classical limit" line.

3. OK, but No Weird Stuff

You recant: OK, you agree that failures of supervenience are not confined to "below the classical limit." But so what? you say. We no longer expect such things to be supervenient. You now admit to being untroubled by these phenomena since they are governed by physical laws. (Remember Mill, who dreamed of the laws of sciences all the way "up.") You aren't at sea with these the way you are with things like consciousness. Laws of nature can fill in where supervenience fails.

Funny, but this is supposed to be *my* point: supervenience fails, but this does not mean that the things in question are lawless. They are still governed by laws—physical laws, biological laws, laws of economics, or what have you. There are laws governing things at all scales of the universe! So, I say, how can it be true that, simply because you are no longer troubled by these failures, since they fall under the sovereignty of some type of order—how can this still be a reason retain of the dogma that the universe is "alive" at only one scale. For the different sovereignties (biological, psychological, etc.) corresponding to different sorts of laws surely signal that the universe is alive at *every* scale.

All right, so this option (3) is not really a way out of an emergentist's philosophical troubles—it is really the way *in* to the scale-free universe. So (3) is a good choice, not a bad one. If you like answer (3), you're on my side. My point is just that it's not an answer that saves the one-scale universe. The position acknowledges that there are substantial failures of supervenience of global on local, no matter where the local/global lines are drawn. It also draws attention to a certain theory that specifically articulates that there is a failure, a theory that is open to failures of this sort at absolutely every level of aggregation—and which is specifically designed to help us think constructively about failures. This theory is capable of telling us just exactly what the consequences of failures look like, and capable too of articulating laws (and therefore predictions) that further advance our understanding of these and related phenomena. It tells us in fact that we must be careful *not* to draw the wrong conclusions from information in one domain, because we might get it wrong—since there is information about the higher scales of aggregation that we have to accommodate. All conclusions about a given scale are in some sense permanently tentative. The quantum theory is therefore an incomparable way of modeling how aggregation itself works, and how to model resheathing. One might even hope that quantum theory is fundamentally a theory *about* resheathings.

We will have further occasion to admire quantum theory, as soon as the next chapter. But now just a final word about supervenience. The lesson at this point is simple: there are failures of supervenience between scales that everyone will consider physical and micro. This suggests that supervenience failures are positively not diagnostic of levels, but they might be diagnostic of new scales of activity. Since quantum theory allows these failures at every scale of measurement, I propose to consider this a *pro tanto* argument for the scale freedom of the universe.

7. Sync Might Be Worse Than You Thought

You might have thought that sync is bad enough, from a metaphysical point of view. And of course you'd be right. But things are actually worse than you thought. I said that the key ingredient in sync models is leakage—that information gets from one place to another, one knows not how. But notice that for such leakage to make a real difference, the information leaked has to be the *right* information—for instance, the right information that allows conservation of angular momentum (say). If all information leaked all the time, that'd be bad enough. But how does a system ensure that the right piece of information gets to the right place and is utilized in just the right way to uphold conservation laws? That is the real trouble. That is why the "physicist" wants local laws, the upholding of which will guarantee the upholding of more global principles like supervenience: the "physicist" wants only the relevant information to reach the relevant destination; and that the information *itself* kick its intended recipient in the ass just so, so as to achieve the desired result. Billiard balls are the model of intelligibility.

The answer to this question, which I hinted at earlier, might actually be worse than the problem of nonlocality—as bad as *that* is. When we look to see how quantum theory models leakage, its own principle seem to speak of the transmission of probability information from one place to another. But how could that possibly work, from a metaphysical point of view? How could it be that the universe knows, at a local level, how to pass along the right probability information so that, in the end, angular momentum is conserved, *with probability 1*? This is perhaps the most puzzling aspect of quantum modeling.

There is a story that Steven Strogatz likes to tell about Christian Huygens, laid up ill one day in 1665. He was working on one of the most important technological problems of his day: determination of longitude at sea. Huygens's strategy (which never resulted in a solution to the navigation problem) employed two independent timekeepers. So as he lay ill in his bedchamber that fateful day, he was kept company by two of the most accurate timekeepers in the world: two pendulum clocks, running to between 80 and 90 pounds of lead each—specimens of the best-known timekeepers, and invented by himself a decade earlier. A quiet day in February, and lo and behold:

> I noticed an admirable effect that no one could ever have thought of. It is that these two clocks hanging next to one another separated by one or two feet keep an agreement so exact that the pendulums always oscillate together without variation. After admiring this for a while, I finally figured out that it occurs through a kind of sympathy: mixing up the swings of the pendulums, I have found that within a half hour they always return to consonance and remain so constantly afterward for as long as I let them go. (Strogatz 2003, 106)

At first he thought that this sympathy was communicated through the air. An experiment conducted with separating wooden planks to prevent passage of air turned up otherwise: the sympathy remained. He had suspended the clocks from hooks nailed into the same wooden support. Now when he turned them at right angles to one another, or separated them by more than six feet, their sympathy disappeared. He suspected they might be communicating through their common support. To test this, he suspended each clock from its own beam, but supported both boards on the backs of two chairs set back to back. The clocks synced. Then Huygens disrupted their synchrony.

> The result must have spooked him—the chairs began to shake. In sympathy they had been motionless, but now they were trembling, clattering on the floor. They continued to shake for another half hour until the sympathy restored itself, at which point the chairs fell silent. (Strogatz 2003, 107)

No matter how the pendulums of each clock started out, eventually they always ended up swinging in exactly opposite directions, like clapping hands. As Strogatz (2003, 108) puts it, "Huygens had discovered inanimate sync." It would be centuries before a single further shred of light would fall on this effect Huygens so admired. Only in the last several decades has significant insight been shed on how inanimate sync comes into being. And that is still not as much insight as we should like to have.

Today we have much more understanding of how sync is orchestrated, as a result of incremental steps taken largely in the twentieth century, with Strogatz taking some of the finest. But still, what we have by way of insight is not so much the *how* of the story, as the *what.* We are learning that, if two pendulums (or oscillators of any kind) are to come into synchrony and stay locked in it, then information at pendulum 1 must continually be utilized by way of "tuning" pendulum 2; and we have a much better idea now of which aspects of pendulum 1's motion are needed, and what kind of effect it is to have if synchrony is to occur. We are also learning a good deal about the conditions under which synchrony can arise and which conditions are prohibitive to synchrony. But we know absolutely nothing about how that information is transmitted, or how pendulum 2 knows what to do with the information it gets. We are no closer to that than before. More importantly, even if we could assume that absolutely all information about pendulum 1 could be communicated (instantaneously, if you like) to pendulum 2, in those instances where we can anticipate the synchrony, we have no idea how pendulum 2 manages to identify the information it needs and "complies" with the demand for synchrony (such as it might be). Here is therefore a place where a couple of nice "local theories" of suitable forces would be very useful. But we have none. Just think about the clattering of those chairs in Huygens's bedroom. That is what propagating information looks like. How do the pendulums filter through all the noise to find just what is required to wash away the clutter? Again: the theories we have tell us how pendulum 2 must respond to certain motions in pendulum 1, and the conditions that have to be in place for that to happen (how far apart the clocks are, for instance). But the question is still: how? We've more insight now, on this side of the twentieth century, than Huygens had on his: and we too just say "sympathy," but we say it with a few more equations and qualifiers.

Pendulums are not the only oscillators with sensitivities. Oscillators that are sufficiently similar tend to "flock." In our time oscillators are of enormous practical importance because we have found that electrons and photons can oscillate. This has led to two technology-enhancing discoveries: lasers, in which different atoms give off light waves that all oscillate in unison, and superconductors, in which pairs of electrons (Cooper pairs, to be precise) oscillate in synchrony, with each pair behaving as though they had the properties of light particles (photons) and thereby permitting electric current to fly through the substance at a resistance of (as far as we can tell, which is quite far indeed) exactly zero. Ordinary oscillators are ubiquitously coupled in nature, showing up, for example, in the synchronized flashing of fireflies, in the chirping of crickets, and in the pacemaker cells that regulate circadian rhythms and heartbeats. So can you rest easy now with the notion of sympathy, qualifiers attached? Can you rest easy *without* an answer to the question of *how* the right information is transmitted and cherry-picked for the right effect?

Now I have to come clean. I laid a trap for you—à la Wittgenstein. I led you down a primrose path, led you to believe that the answer to the hard question I kept badgering you with had to lie in a principle of locality, which led you to believe that the answers to all such questions lie in thinking about how information travels—at the most minute scale—to reach a certain place where it is utilized "locally." Posing the question is honest work. But more honest still is the proposal I now make to you: it is philosophically respectable to reject the question. So I urge you to reject the suggestion, unstated but implicit in the way I stated the problem, that this "locality" principle is the only way to think about the "action" of the universe. Here is how: failing to find a locality principle that can do the job, we might consider that the very notion of scale is not what we think it is—that scale is conceptually more complex than it seems, and so action of high-scale characters upon characters at a lower scale is not as problematic as it seems. Here is what I mean.

If you live at the scale of a gnat, then certain things are "present" for you, and certain other things are "remote": "present" and "remote" are scale-relative attributions, because larger entities have more things "present" to them—their scope is greater. Adherence to locality principles tends to obscure this fact. Consider the possibility that not all the activity of the universe occurs at the scale of a gnat. Then at one of these higher scales of activity, what the gnat will view as remote will instead count as present. And so it will amount to being local *in a different way* (as different as the difference between an *intensive* variable and an *extensive* one, which comes up again in a parenthetical note in the next section); it's as if the information needed were simply there, and by that I mean everywhere within a region of determinate size scale, and so were apt for interaction or utilization on the turn of a dime. What's remote at one scale is local at another and so can act there even against lower-scale prohibitions (if such exists) against its acting there. No traveling required. End of story.[13] That is what it means for the universe to be scale free.

The suggestions of motion or oozing, in my earlier usage of the term "leakage," was therefore misleading you via a suggestion that the answer must lie in local activity—where local meant "local at the level of micro"; therefore you never thought to inquire whether the relevant activity might be transpiring on a larger scale: entangled states becoming disentangled, for example—just like that. No traveling or communication at lower scales required. The energies involved may not be the energies of things communicated and things kicked from here to there, but may be of things finding collectively states of greater organization or lesser total energy—the apparently many acting as one.

You can resist my new suggestion of scale-relative locality: you may make no sense of it as a true conception of activity, or even of locality—you may not see it as anything but language covering over a gaping hole, a hole that cries out for the discovery of activity at an appropriately micro level of activity. OK, I reply: good luck finding that.

Still, my answer to the "hard," but now I'm saying rejectable, question, which some minds may nonetheless still find attractive, might be unpalatable enough to make you wish you had a different model of how supervenience fails. That is where our discussion will go next.

8. Networks: Categorical Connection Models (*Linked* as Contrasted with Synced)

> [T]he state is by nature clearly prior to the family and to the individual, since the whole is of necessity prior to the part; for example, if the whole body be destroyed, there will be no foot or hand, except in an equivocal sense, as we might speak of a stone hand; for when destroyed the hand will be no better than that. But things are defined by their working and power; and we ought not to say that they are the same when they no longer have their proper quality, but only that they have the same name.
>
> —Aristotle, *Politics*

[13] Here is how Chester (2003) characterizes the action of "measurement" in quantum theory:

> Before the measurement there is no particle pair; there is only a gigantic atom. This atom pervades all space. The experiment dematerializes the atom, and in its place two particles appear. Each materializes, as it must in the universe, so as to preserve the laws of nature.

Like ferromagnetism and the structure of matter, which we discuss at some length in chapter 2, superconductivity needs to be explained by quantum principles. Superconductivity was discovered a century ago by Heike Kamerlingh Onnes, who was messing around with very low temperatures. The story is suspensefully told by Strogatz in *Sync* (2003, 128ff). I'll bore you with just a few details narrated in my flat-footed way, stressing heavily the philosophical points that are so lightly touched on by Strogatz (but urge you to read it in his own delightful narrative). The story of superconductivity begins with a crucial fact about entities that only quantum physics (among the natural sciences) seems to recognize at this point in time: the fact that ensembles of entities have characters. Ensembles of electrons, for example, belong to a category with so-called Fermi statistics (a certain way of counting things)—they are *fermions*; while ensembles of photons follow Bose-Einstein statistics instead—they are *bosons*. We can't know what class an entity falls into unless we study the entity in its throngs and scrutinize their behavior in groups. This (among other things of course) is what we find when we do: fermions are "shy"; they never crowd each other; their statistical properties exemplify a kind of "spread out as much as possible" character—distributing themselves among the energy states they are allowed (by other constraints) to occupy. Bosons, by contrast, are gregarious; the more of them in a room, the more others will join—they are partiers. These characters are fundamentally important to understanding how entities behave when you get a number of them in the same room. This idea was crystallized by Einstein in 1924, who attributed its origins to an inspiring letter he received from a young Indian physicist named Satyendranath Bose. Because of their gregarious character, bosons enjoy a very high level of quantum sympathy: they act very strongly as one. Einstein proved that, at the right temperature, bosons fuse and lose their identities, acting literally like one entity. It's a new and ineffable kind of matter, called a Bose-Einstein condensate. Our first lab encounter with it was in 1995 in a high-tech lab in Boulder, Colorado, a rendezvous for which the hosts received the Nobel Prize. This "merging" capability (as we'll discuss in the next chapter) is characteristic of waves, less so of particles.

The whole trouble with superconductivity is this: when it comes to superconductivity, we're not even talking about bosons. Superconduction occurs in the electrons (fermions) of materials that, at room temperatures, do not exhibit even the most modest amount of ferromagnetism, something typical of a good conductor. It's a defining feature of superconductors that they are *not* good conductors in the ordinary sense at all. And superconductors really are *super*conductors: their resistance to motion of current is, to an incredibly high degree of precision, *zero*. This would make them incredibly useful in the power grid, as we'd be able to kill many birds with one stone: conserve energy, eliminate waste, and reduce carbon emissions. If only we could employ superconductors! Unfortunately, that is a long way off because we still do not have materials that are superconductors at anything close to common air temperatures.

In superconductivity, electrons seem to be piling, in the fashion of bosons, into the same energy level—the lowest possible for their external constraints—and moving together, it would seem, as though the many were really just one, and so acting as though there were no electrical resistance at all. In this they are violating all their personal characters as fermions—they've become sheathed, like the fabled wolves in sheep's clothing. How does this happen? The key idea, due to John Bardeen, Leon Cooper, and Robert Schrieffer (BCS, as they are known for short) was formulated in 1957 (1957a, 1957b; drawing also on Cooper 1956). The basic idea is that, when submerged in a sea of opposite-charged particles (and we'll discuss the concept of such seas in the next chapter), electrons form pairs. But a Cooper pair of electrons (separately fermions) is a single boson! Talk about resheathing! This is an excellent example of what I have in mind by a higher-scale character (in this case, the oppositely charged ion sea) acting upon something at a lower scale (the two electrons, fermions) to create of them something completely different. A Cooper pair is much more like a quotation than it is like an original statement. Within the shroud of a Cooper pair, a couple of fermions act as if they are a single boson, in sync with all the other bosons and dancing to the same superconducting tune.

Cooper pairing requires an appropriate context—in the superconductor the context is the sea of positive charges (a lattice of nuclei) of the superconducting material. It is this context that makes the pairing possible. Here's a close (but imperfect) analogy. Think of the lattice of the superconducting material as a structure like a stretched rubber sheet, or a taut drum. Electrons circulating in the lattice will be like small marbles on the stretched membrane: each one deforming the membrane just a little, drawing more of the membrane around itself, thereby creating a small pool of excess positive charge around it. As this happens, its nearest electron neighbor is drawn to it. Pairs form (rather than triplets or other multiplets) on the BCS model—for reasons having to do with relative energies of different configurations. At a specified temperature, we can get a condensation of the entire sea of electrons into Cooper pairs. The whole business must take place at a low enough temperature for the attractions between the small poolings of excess charge to swamp other effects.

The stretched-membrane analogy isn't quite right: in the BCS theory, the electrons don't interact with charged lattice ions so much as with "phonons"—disturbances or waves on the entire lattice structure of charged nuclei, conceived as a single entity or medium (something we will discuss in the next chapter). And the pairing of electrons results, not so much from the attractions of charges, as from their wave functions overlapping—a fact with significant consequences only at the low energies we are talking about. In other words, the effect is profoundly quantum mechanical in the sense that it involves overlapping of probability waves; this requires a context in which that overlapping is dominant.

Now, the membrane analogy could have been drawn somewhat differently, as follows—though this articulation will *not* be in line with the BCS theory. Still, it is suggestive of how one can tell quantum mechanical stories; this is in "conceptual sync" with the spirit of explanation I am trying to evoke. So imagine now the electrons as droplets of water, initially randomly distributed over the membrane. Then imagine that each individual electron "pool" vibrates as the membrane is slightly disturbed. This makes it spread out a little bit. Because the entire membrane is vibrating, eventually all the water will be evenly distributed over the membrane and vibrating in sync. In that situation, it would make sense to speak of there being only one mass of water, acting as a single, strangely quivering "droplet." (As far as I know, there is no kosher mathematical model, utilizing physical principles, that has an appropriate correspondence, or even a resonance, with this fanciful analogy. And even if there were one, I make no claims that it would be as well corroborated by predictions of specific heats and other effects as is the BCS. This is simply another analogy with which to think.)

You might not like either of these models. Both might be too "quantum" for your taste, involving too much nonlocality and loss of identity. You might be favorable therefore to developing a different, potentially more metaphysically respectable model to handle the synchrony we cannot deny exists. You might want a model with more obvious local conservation of information "seepage"—a model with more obvious information flow characteristics. Beware: there are theorems to the effect that no such models are compatible with a range of quantum effects. (The first and most important of these is Bell's Theorem, mentioned above, to roughly the effect that no physical theory of local "hidden" variables can reproduce all of the predictions of quantum mechanics.) But you might not be interested in capturing *absolutely all* quantum effects with your model. Still, you should reread the first half of this chapter to make sure you are capturing the ones you intend, compatible with your position on supervenience.

In any case, it might be worthwhile looking at a different model of coordination, simply for the contrast. We'll try for something of this sort now. A word of caution: these models will prevent violations of supervenience, but they come with their own kinds of complexities, as we will find, as we launch now into characterizing them.

Rather than a "sea" of materials, one could instead conceptualize entities as forming a categorical network or lattice framework. And rather than utilizing the math of quantum mechanics, one might instead use the mathematics of what has been called familiarly *artificial life* (Alife) or *cellular automata*, but the better labels are *homogeneous structures* or *iterative arrays*. This math is much, much simpler, as is its interpretation; and its aggregation is the aggregation achieved through the confluence of computer code rather than the "integrative methodology" of (differential) equations. I will refer to models of this sort as *linked,* in honor of Lazlo Barabási's elegant 2003 book of the same name.

The basic idea is to construct a computational model, utilizing a regular grid of "cells," whose properties and steplike dynamical transitions are simple and dictated to it, and which conclude each steplike transition in a definite (classical) state, such as 1 or 0, on or off. For each cell, a set of cells called its *neighborhood* will be clearly defined, and these will be the cells with "influence" upon it. For example, the neighborhood of a cell might be defined as the set of cells a distance of one cell unit or less away. And suppose that the "new generation" of cells is generated by this simple rule: a cell in the new generation will be on in the succeeding generation if exactly two of its neighbor cells are on in the current generation, and otherwise it will be off. Routinely, the rule for updating to the next generation is the same for each cell, does not change over time, and is applied to the whole grid simultaneously.

We will discuss a model of this sort in chapter 3 that seeks to handle the staggering flocking behavior of many species. You don't need much information to follow the rules if you're a bird in the flock and what information you need is local. (For example: avoid collisions with nearest neighbors.)

For the nonce we wish to focus on the *network* properties—the properties explicitly assigned to the nodes. Idiosyncratic features of nodes are explicitly treated as totally irrelevant. The arrays are truly homogeneous. They are, in fact, simply *fiat* characters: they are had because we say they are had, and we have no independent reasons for thinking the characters are what we say they are. In a network model, we simply decide which properties are network-relevant and build models that look at an arbitrary spatial aggregation of these; we don't make any effort at all to "figure out" how these characters "emerge" from the characters of node occupiers.[14]

Models like this one can do a very nice job of explaining such things as flocking. To the extent that we wish to maintain that they are successful, flocking should be characterized as *linked* rather than as *synced*. The point I wish to make now is that there is quite a smooth transition between linked and synced models.

It is easy to see how one can convert a simple linked model into a synced model. All one has to do is to allow what is called "coupling"—allow the states of cells to depend upon the states of other cells in the same generation. And immediately we get failure of supervenience. If that coupling is over any appreciable distance—for instance a distance one order or magnitude larger than the size of the neighborhood, we immediately have violations of locality. The typical linked

[14] Epstein (2011) makes similar observations about agent-based models and concludes that one cannot draw familiar locality conclusions from them.

model is rife with violations of local conservation of seepage. In linked models, the seepage is clear, as is how it happens. Linked models involving coupling of cells, in real-life physics, are known as *coupled map lattices*, and condensed matter physics is littered with them. We will eventually take these up again (in chapter 8), at much greater length.

Linked models are transparent: everything about them is dictated by the researcher. Some features stand out as unrealistic: inviolable grid structure is one, categorically imposed. There are no properties at the nodes except those that are explicitly assigned by the mathematics. You might be prepared to accept these features as simple idealizations, to achieve a model that is genuinely homogeneous. But isn't homogeneity itself a property that a real-life aggregate might or might not enjoy? And how does a departure from homogeneity affect the results? That's really very difficult to study within this framework. This is something to keep in mind when assessing the value of each model (an assessment we conduct to some extent in chapter 3).

Now as to the question: is it possible to model the failures of supervenience in quantum mechanics in a linked (as contrasted with a synced) model? The answer is yes, at least in regard to the character (spin) we looked at above, so long as we can manage to build in the relevant coordinations. For example, we can impose a simple rule: in a population of paired cells, a cell in the new generation will be Up in the succeeding generation exactly if its "partner" is Down, and Down if its partner is Up, and otherwise will have some third (neutral) value. (We will have to impose other rules, obviously, to reproduce the half-Down/half-Up statistics of the entire experimental setup, but that's not of concern at the moment.)

So now it's clear how to get violations of supervenience. (Of course we've said nothing about how it must be done *on the ground or in the lab*; just theoretically how possibly it might come about in a lattice context.) It looks like it helps to have some kind of rule, "bred in the bone" of node inhabitants, as it were—some preprogramming. But there's nothing special about having *shared* preprogramming; it might be possible to have the same level of supervenience failure with exactly *half* the preprogramming—so long as *some* preprogramming creates coordination, the *other* preprogramming (if there is any) can be anything at all. And it helps to have seepage—to have information flow.

Because linked models are more transparent, failures of supervenience can be pinpointed (in the sense that we can say what features of the model are responsible for them). Does that make these failures any more palatable? Well, not really. If you were not bothered by failures of supervenience before, you might actually be *more* bothered now. Why? Because these models make global features of the objects they mean to depict much more diffuse, and so, much more difficult to keep track of. Here's what I mean.

Recall what I said at the end of the last section, by way of trying to reconcile you to supervenience failures: I said that there were more features "present" at any given point in space than you were keeping track of—namely, the ones at the higher scale. These could act at their own (higher) scale, upon the features at a lower scale. Now the trouble with a linked model is that it's harder to find those global features—because they are not defined, but quite literally require making computations to work out. That is the nature of linked modeling. Still, these global features always exist. Let's see whether they provide the required help—whether they allow us to forestall the need for sync modeling.

What counts as a higher-scale property in our impromptu "paired cells" model of the experiment in the lab? Obviously, the joint trait: Down/Up and Up/Down are its values. Obviously also if we don't measure, we don't have values (that's what quantum mechanics prescribes), but that too is a matched pair of no-values. Or we can have Down/no value, no value/Down, and so on. That is a trait worth keeping track of. I said before that it is this quality that we have to thank for (say) the outcome at the B wing, when once we have measured the one at the A wing and found it determinately Up or Down. It is because the joint trait is Up/Down that we now anticipate a Down at the B wing when once we have found it to be Up at A.

But now someone with a distinct preference for linked models will say: no, it is because the cells are observing the imperative "anticoordinate!" that we observe the obverse result at B, when once we have observed Up at A. So the models are giving *distinct* explanations of the phenomena. The trouble, of course, is that now the linked model has to explain how it is possible for a cell to observe the directive "anticoordinate," when the cell itself has no trait at all. So if the models are genuinely different, it is the linked model that is the less palatable to interpreters with sympathies for a one-scale universe.

But in fact it would actually be a mistake to characterize the linked model as depicting a one-scale world. In fact, lattice structures end up with traits of their own that do not supervene on the traits of the cells, and which have profound effects on the performance of the lattice—for instance as a data network.

Consider the case of a network of just nine nodes (three by three) as in figure 1.2. This network forms a connected graph: there is a pathway from every node to every other. The average length of the shortest path between any two nodes (or the average *geodesic distance*) is roughly 4. Figure 1.3 is also a connected network, but the average geodesic distance in this network is somewhere between 1 and 2. Because of these different geodesic distances, the two networks will have different properties (average characteristic time for travel of a message from one "end" of the network to the other). And the two networks will have very different fragility properties (susceptibility to "crash") from the one in figure 1.4, which is *not* a connected graph. So the property of being a connected graph is tied (causally, if you

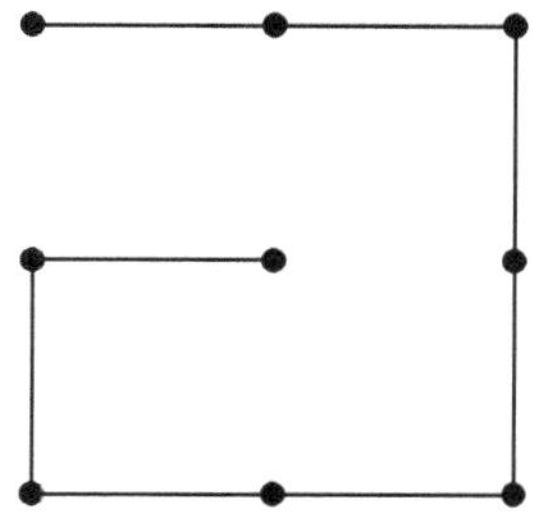

Figure 1.2 Connected graph: the average shortest distance between nodes is 3–4.

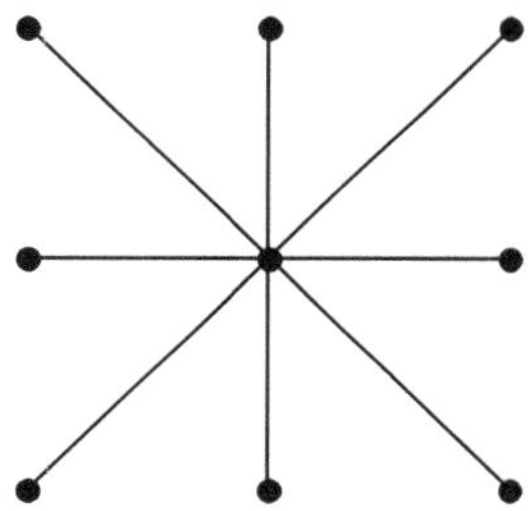

Figure 1.3 Connected graph: the average shortest distance between nodes is 1–2.

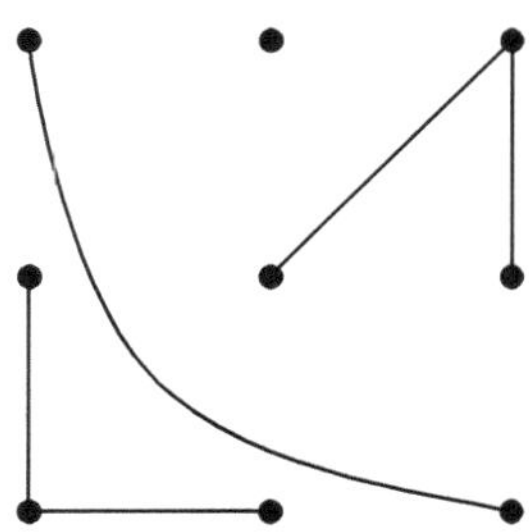

Figure 1.4 An unconnected graph

like, though I will explain in a chapter 4 why you should not like this language) to other macro properties of the networks.

Now, is the property of being connected supervenient on the local properties of the nodes? Consider just the property of a node defined as the number of other nodes to which it connects (called its *degree*). You can't recover the property of connectedness even if you knew the degrees of all the nodes—the degree information typically underdetermines the full graph. How about the property of being connected to one's northeastern neighbor? You can recover the global characteristics of the network from such information *only* if you also have relative position data (like "connected to northeastern neighbor") on each and every node—in other words, if you know exactly, for any two nodes, whether they are connected by an edge. And that means you need to have a great deal of knowledge about the *spatial* global structure of the network in order to recover its connectedness properties. Is this a violation of supervenience? Perhaps not. But consider: in order to determine whether some graph is connected, you need to know, not only information about where all the nodes are situated relative to one another, but also precisely how large the network is. And this would seem to be global information (in the language physical chemists use, it requires utilizing an extensive variable, not just an intensive one). So the question becomes: is there a violation of supervenience in imposing the "local" rule on a given cell (for example the rule: "Spin Up if you are part of a connected graph")?

If the slogan for supervenience is that we can recover global information from local information, wouldn't this case seem to be a violation of supervenience? (The character of connectedness in a system implementing the rule we have been examining, "Spin Up if you are part of a connected graph," would then be self-referencing—which in this case is problematic, because it would cause the computation to be nonhalting; we would end up with a noncomputable function on our hands.) Surely network architecture is global information if anything is. And it would seem to be true that no specification of network architecture ("spec" as they say) can be recovered from any-size aggregation of local information. If that's true, then certain features of networks cannot be referenced in the "localized" decision rules employed by cells. If these are required to explain the behavior of the network, then network behavior is not explicable in terms of local behavior.[15]

9. Models: A Retrospective and a New Outlook

Scientific modeling is a diverse enterprise. Still, there have been numerous philosophical programs intent upon articulating what a scientific model is—or anyway what it ought to be. One of the most eligible and irrepressible ideas in this

[15] Logicians have a version of the failures of supervenience of global upon local, of general upon specific, of quantified upon atomic. This is like the difference between an arbitrarily long conjunction of atomic statements of the form "Object i is P" with "i" running through the subscripts so as to exhaust all the objects in one's universe, and the statement "All objects are P": the latter is not deducible from the former. In laboratory contexts, especially in chemistry, it is a commonplace that no extensive variable can be recovered from any number you like of intensive variables.

area is that a scientific model has to be roughly causal, and that it must supply a mechanism that essentially plays the role of "decoding" the unintelligible in terms of something more intelligible—the mechanism. It should come as no surprise therefore that the notion of mechanism is often accompanied by the idea of local forces (which we encountered with Davies earlier). Newton supplied one of the most vivid examples of scientific modeling via forces, though his were truly nonlocal! (A gravitational field whose source is itself quite local—a certain body, for instance—could act simultaneously at all the points of space.) But what he did was to display so splendidly the power of a single idea that explained the motions of planets, tides, and apples—so much with so little.

Pushes and pulls. That *was* science. No so much any more. At least, not so much in Physics. In Physics, knowing all the forces at work in a given system is just the beginning. We know now that a physical model must do much more to be genuinely adequate. And perhaps surprisingly, we still don't have a general theory of matter in all its forms. In my estimation that is the most profound of all open problems in Physics. We need to explain how forces are themselves the result of the structure of matter—and how matter itself comes about, and how matter and space are related. Forces, as such, do so much less than they used to do.

The reason is simple: the world is really not governed by Newton's laws of motion—Newton's dynamics; the basic laws of the universe are not laws about how (inert) matter responds to the creative activities of forces. That narrative is at best incomplete—however useful for sending rockets to the moon or playing basketball it might still be. But what is the complete narrative? What, indeed, is the quantum narrative? This is still very much a work in progress. But I want to bring out one aspect of it as we move into the next chapter: the wave/particle duality and what this adds to the narrative.

Notwithstanding the difficulties we do have in understanding the quantum narrative, what is abundantly clear is that we know how to manipulate useful formulas under the guidance of some principles of analysis we get from Newtonian physics, but also under the guidance of analogy with principles of optics and wave physics. Revealingly, we can transport one set of principles from one context (a context of waves in a medium, say) to another context (one with particles only) and articulate analogous reasoning in the new context, whenever one set of equations suggest another. And this works surprisingly well indeed. (It is how the Schrödinger wave equation was itself discovered.) Where this transpires, we often get head-scratching predictions—that pan out!

But the principles underlying wave analysis are not so different from those that underlie particle analysis. And this is really perhaps the deeper moral. Both sets of principles result in (for example) principles of conservation (energy, momentum, spin, etc). Still, particle physics contains a surplus of principles, certain principles that wave physics does not seem to have counterparts of: the principle of locality, for instance. These are principles of restriction or prohibition: they are restrictive in ways that wave physics isn't restrictive. The same is true also in reverse: wave physics involves prohibitions that particle physics does not—the principles of energy-level occupation are instances of this latter reality. Let's focus on these for the moment.

The Pauli Exclusion Principle dictates that no two electrons can occupy the same energy state in the same quantum system. So they have to "stack up" into the higher levels when there are more of them. (And that's why they are less social than their boson colleagues.) Perhaps that is why matter here on earth actually occupies space; Freeman Dyson and Andrew Lenard present such an argument (1967, 1968). But principles of this sort don't come out of Newtonian physics of matter. So there are important differences. (I'll talk more about differences between particle-based mechanics and wave-based mechanics in the next chapter.)

The upshot of this is that we cannot always rely on a classical picture or model to explain a process—and indeed that classical pictures can be very misleading. One has to study the principles involved and come up with models that are faithful to them. My favorite example of such a model is one due once again to Steven Strogatz: watermelon physics. He uses it to explain how a laser works.

Imagine that you wake up one day to find yourself on an alien planet. The only thing you see for miles around is a watermelon, and beside it what looks like a companion stool just the size and shape to accommodate the watermelon. Curious, you set the watermelon on its stool. And something somewhat surprising happens: the watermelon jiggles as if agitated for a while, and presently it jumps off its stool and back to its original spot on the ground; at the same time, it shoots out a seed like a bullet, in a random direction. (You check it out several times and discover that the watermelon doesn't seem to have a direction preference when it comes to spitting out its seed.) For the rest of the story, I will just quote Strogatz (2003, 111) at length; I can't articulate it better myself:

> As you continue to explore the planet of the watermelons, you soon come to the edge of a vast field, with millions of watermelons lying on the ground, each with its own stool next to it. This raises an intriguing new possibility. What would happen, you wonder, if a seed-bullet happened to strike another watermelon? To start the action, you lift one melon onto its stool. It soon falls off and fires a seed in a random direction, and by good luck, its flight path carries it smack into another melon lying on the ground. As soon as the target melon absorbs the impact, it jumps up onto its own stool, where it too starts to quiver, and before long, it drops down and fires a seed of its own, again in some random direction. It's an amazing spectacle, one seed triggering another, melons jumping up onto stools and then falling back down. By

> lifting the initial watermelon, you have inadvertently created a chain reaction, although of a very feeble, nonexplosive sort: its size stays fixed at just one seed in the air at all times. Actually, if a seed ever fails to hit a watermelon and flies out of the field, the process stops altogether.
>
> This cascade process is interesting, but it's not the analog of a laser. It doesn't amplify light; it never increases the number of photons in the air. The missing piece has to do with the only aspect of watermelon physics we haven't considered yet: What happens if a seed hits a watermelon while it's tottering on its stool, instead of one lying calmly on the ground? To find out, you lift many watermelons onto their stools at the same time, running quickly from melon to melon before any of them has a chance to roll off. Then you stand back and watch. Eventually a melon drops down spontaneously and fires a seed, scoring a direct hit on another melon wobbling on its stool. (The odds of this are good, because you've lifted so many melons onto their stools ahead of time.) What happens next is astonishing. Instead of being absorbed, the incoming seed passes straight through the melon without changing its flight path; what's even weirder, it is now accompanied by another seed exactly like itself, moving in tandem with it. In effect, the incoming seed has been cloned. Before, there was one seed flying in that direction; now, there are two.

The cloning process is the key to laser phenomena. It is a quantum radiation phenomenon—spontaneous emission of energy in the form of photons with certain features.

Strogatz's model assembles in one place all the ingredients required to explain the important features (in this case of lasers) that we need explained. And so, however fanciful, watermelon physics is an excellent model of laser technology. It's no more and no less "alien" than the principles encoded in the companion mathematics, but it has the advantage of visuality. It makes very stark what we don't find intuitive about certain features of the laser phenomenon. (It's fantastical for that very reason.) To the extent that models help us—even when they violate our "common sense"—we need to embrace them. And if that means giving up some cherished principle, so be it. Common sense be damned. Or at any rate, common sense be retrained! That's why science training is *education*.

What is the result of retraining? Does it give us the "mind of God"? Stephen Hawking is famous for having concluded his bestseller *A Brief History of Time* (1988) with a credo: "If we do discover a complete theory [a Theory of Everything, in the terms of the introduction to this book], it would be the ultimate triumph of human reason—for then we would truly know the mind of God." What he meant by that, of course, is that we should understand the fundamental principles of the universe. I *don't* think he meant that our common sense (our natural understanding of things) requires retraining—a point I am trying to urge upon you. I wager that we won't be getting anything like the TOE that Hawking envisioned. So I am urging that we need models like watermelon physics—a different watermelon physics for every occasion—to improve our understanding: with watermelon physics firmly in mind, we can be rightly said to understand something about the universe we failed to appreciate before we had watermelon physics (or anything like it). But this something that we are appreciating, via a grasp of the (relevant) watermelon model, is *not* more important than any other something that there is to appreciate about the universe—it is *not* something more true; it is *not* something more fundamental. The question of fundamentality is a philosophical question to which is devoted the entirety of chapter 3, and then some.

Still, the something that we learn when we grasp a certain watermelon physics can be leveraged within the context of the unity of science. For the principles behind the model might be useful in other areas of inquiry. Science, like the world itself, is *modular*—it exhibits the articulations into rough areas of apparent autonomy. To the extent that principles *port*—to the extent that models like the watermelon physics model can provide guidance in modeling other phenomena as well—to that same extent are the boundaries between different disciplines *porous*. And if we are able to carry principles of watermelon physics over into other modeling contexts successfully, that will be abundant proof that the principle of unity of science is of singular importance to the philosophy of science. And proof there is: like so many physical models, watermelon physics integrates principles of modeling from a number of other physics contexts.

Now a general language that describes the intercommunication of models is required, and required also to be as precise as possible. It will aid us in seeing how science is united, and furthermore it will advance the cause of model sharing. So philosophical articulation of a language that helps to make models more transparent to researchers in different areas—and so easier to analogize in new contexts—that should be the goal of the philosophy of science. It is my goal in this book. With that goal in mind, I invite you to continue the journey into chapter 2.

2

Resheathing

somebody almost walked off wid alla my stuff
not my poems or a dance i gave up in the street
but somebody almost walked off wid alla my stuff
. . . .
this is mine / ntozake "her own things"
that's my name / now give me my stuff
i see ya hidin my laugh
& how I sit wif my legs open sometimes to give my crotch some sunlight
& their goes my love my toes my chewed up fingernails
. . . .
i want my stuff back
my rhytums & my voice . . .
. . . .
who is this you left me wit
some simple bitch widda bad attitude
. . . .
i wants my things
i want my arm wit the hot iron scar
& my leg wit the flea bite
i want my calloused feet & quik language back in my mouth

—Ntozake Shange, "Lady in Green," in *For Colored Girls Who Have Considered Suicide When the Rainbow Is Enuf* (1975)

1. The Quest for Stuff

Ntozake Shange's powerful choreopoem *For Colored Girls Who Have Considered Suicide When the Rainbow is Enuf* was first performed in the late 1970s. According to some, it is the black feminist's *Hamlet*. Lady in Green's soliloquy comes toward the end. In this marvelous poem, entity and quality merge and separate. We wonder at first what "stuff" was taken from Lady in Green, what items of a personal nature she might be feeling naked without. But soon we realize that she is talking about her own very qualities, the characters and traits that make her what she is distinctively as a human being; they are her "stuff." And the "somebody" who precipitated her diminishment (from a delightful self to a regrettable "simple bitch widda bad attitude," like a pair of electrons wearing the mantle of a single boson) is portrayed simply as a petty thief. Lady in Green blurs magnificently the line between object and quality, flirting strikingly with analogies physicists draw on all the time nowadays. Poet and physicist each leaves us with questions about how "stuff" gets to be "ours," how we manage to get the (good) stuff in the first place and, perhaps most profoundly, how the world manages the conservations of certain stuffs. (Indeed Shange's poem manages a sort of glimmer of an answer, involving the sisterhood of women, a form of sympathy that Huygens himself might have admired as much as the synchrony of clocks.)

The poet behind the Lady in Green is enjoying a liberty quantum physicists take for granted every single day: the liberty to carry over the physics of waves, and the principles that seem to apply only to waves, into treatment of particles—and then to turn right around and do the same again but in reverse: carry principles that apply to particles over into the theoretical analysis of what seem be purely wave phenomena. Physicists now routinely take a property of a "wave" solution to a differential equation (such as the Schrödinger equation) and treat it simply as a thing—as stuff. That's basically how quantum physicists were able to articulate a theory of atomic structure—a profoundly important achievement in twentieth-century science. This chapter is devoted to discussing this "play," and examining some of its philosophical implications. This will give us some insight into the nature of stuff.

Poetry, like physics, flirts with and then subverts commonsense metaphysics. On display in Shange's poem is the wave-particle duality, in all its glory. And taking one step back from the Lady in Green to the very genre of narrative performance (moving therefore to a somewhat larger scale), we encounter another means of the subversion of common sense. Theater itself (a "larger form" of poetry in a performance setting) is a kind of wave, in that it embraces certain "harmonious ambiguities," even to cultivate them—in contrast to the more literal-minded, more "particulate" characteristics of prose. Consider for instance the soliloquy, the speech that ensues when a character turns away from the scene to address the audience. The transition is a jarring "move," a departure from certain conventions and proprieties that are otherwise solemnly observed throughout a work (or they'd be much less powerful when broken), to a different form of communication; the soliloquy is a "singularity," a dimensional shift. The audience knows without having to be told that the author is changing rules, changing the "scale" of the communication. The audience feels the grip and grind of the transition, as they adjust to take in the message. If the audience does not experience the transition, the message cannot be communicated; the jolt is part of the message—or at any rate, part of what it takes to send a different kind of message across.

Off-stage, however, there is less grinding and transitioning. Poetry is a singular experience of communication within the vast expanse of ways that humans communicate. Change of scale is not a commonplace in that larger setting. Activity within-scale, more or less, is the rule. This is far from saying that the universe is single-scaled; to the contrary, it must be scale free if beings like ourselves—large, slow-moving, mostly oblivious of the lightness and versatility that is the quantum—are to be in a position to act in it at all. And act we do; we make stuff at our own scale, more or less. We do it by resheathing old stuff in new skins, at roughly the scale we ourselves inhabit. And that will be the second important set of insights we will be after in this chapter.

2. Mechanics

Mechanics as we know it today, as an aspect of Physics, originated with the modern (roughly Enlightenment-era) assumption that a body is simply an unstructured assemblage of quantities that attach (how is unspecified) to one another, thereby forming a body.[1] In the mechanics of our present era there is no attempt to account for how one body differs from another, as a body. The mystery of body is unexplored—a pity, since (as I'll discuss in the conclusion to this book) the mystery of body is no less deep than that of mind. Therefore there is no provision in physical science for how to count bodies, even once we have produced an inventory of all the quantities or the features making an appearance within a certain expanse of spacetime. The central mission of mechanics, in the contemporary scientific enterprise, is instead to secure an accurate account of the impact of time, by setting down explicitly how things change over time once everything else (namely the quantities and other features of the system that are at least in principle measurable) has been specified for some interval before. I will refer to this body of theory as *dynamics*.

There are two formulations of dynamics, the first linked to the name of Newton and the second linked to the names of William Hamilton and Joseph Jacobi.[2] The differences between them are monumentally important. We will examine this difference, by way of entry into the topic of waves and particles.

[1] A quantity is a feature that may vary in *magnitude* with time, taking on no more than a single magnitude at a single moment in time. Thus a quantity is a concretum, metaphysically speaking, which possesses an identity through time; it is not an abstract object like the mathematician's variable.

[2] For a full treatment of this subject see Thalos (1999a and 1999b).

Now a theory of dynamics requires embedding in a general theory of quantities and their potential dependence relations because dynamics alone is not sufficient to produce a thoroughly general treatment of motion and change over time. Specifically, dynamics does not contain a systematic treatment of the nature of quantities as such, nor a taxonomy of quantities, nor any general account of how different species of quantities are related, both as time goes by, and at any given instant in time. Such a treatment would first have to answer questions like these: How are macroquantities related to microquantities? How, precisely, is macro dependent upon micro? When in fact is there dependence? Does dependence always look the same, hold between the same kinds of things? Does dependence itself change over time? Can the macro play any of the same roles in mechanics that the micro plays? Once furnished, such an account of quantities would provide the means, at least in principle, whereby to articulate complete descriptions of systems comprising enormous numbers of interacting parts, with proportional numbers of interacting quantities.

As we've already seen, these questions are profoundly important for understanding how the universe operates; and the answers we incline to give them, off the cuff, are highly likely to be wrong. With only intuitions to rely on, we're very likely to get it wrong as to whether the universe operates at one or multiple scales; our evidence before formal experiments are conducted is so meager, our intuitions baseless. So we should be prepared to revoke assumptions.

Consider the porcelain cup I am drinking from right now. What shall we say is the relationship between its microscopic realities—the to-ings and fro-ings among molecules that constitute it—and the fact that the cup shatters upon impact with my kitchen floor? Is the relationship a one-way affair, in which the microscopic gets absolutely all the credit for what transpires at larger scales? If we answer yes, we are endorsing a hypothesis to the effect that the microscopic is—always and everywhere—master, and that the macroscopic is always and everywhere slave. It's the one-scale model of the universe. I have argued in other places that this master-slave doctrine, which in this book I am calling the *one-scale model*—is the cornerstone of a certain Newtonian image of the world, characteristic of eighteenth-century mechanical theory. Bodies in the world, according to that theory, form familiar middle- and macro-sized objects—where they do—because atomic particles interact via forces; these forces, together with the motions to which they give rise, are entirely responsible for the macro features, such as for example the mechanical and thermodynamic qualities that we measure at macro scales. Isaac Newton contributed to this conception of things—though as we saw in the introduction, he could find no reason to impose an upper bound on the size of fundamental particles. Still, the mechanical vision of the universe is in large measure due to the analytical principles Newton exemplified in his work, so I shall refer to this mechanical vision by Newton's name, though we shall do well to bear in mind that he might not have embraced it in full.

According to the Newtonian vision, it is indeed axiomatic that macro effects are brought about by macro causes, but entirely in virtue of the fact that micro causes bring about micro effects. For example, my cup shatters because certain intermolecular forces in it (as a porcelain article) are in very fine and quantifiable ways different from those in, say, a steel fork or a textile, and are such as to make the cup (in common parlance) "fragile." According to this vision, the macro has no independent causal reality, because the universe is active at only one scale. And this way of putting the Newtonian doctrine brings into sharp relief the question of dependence relations among quantities.

Newtonian Mechanics

A natural description of motion introduces *vector* quantities—quantities with magnitudes, anchored tails, and orientations—to portray them, much the way we might portray different-sized arrows. (And it should make us alert to the question whether all quantities will be "localizable"; sadly, the two questions of [1] how to introduce quantities and [2] whether quantities are in general localizable are rarely if ever discussed in relation to one another; I know of nowhere where they are discussed even in proximity of one another.) The Newtonian formulation of general mechanics[3]—historically the first comprehensive formulation—takes vector analysis one step further, for with each alteration in motion it postulates at least one entity—a vehicle or emissary, routinely interpreted as a cause—that brings about the alteration by communicating an influence in the appropriate direction. Like ordinary bodies, these vehicles through which influence is communicated are both located in spacetime and propagated along continuous spatiotemporal paths.[4] The distinguishing marks of the Newtonian treatment are then as follows:

1. Influence propagates from agent to patient, via a vehicle. This is so (according to the Newtonian conception) because body does not act upon body directly, but only through an intermediary. The vehicle is thought of as originating from the agent as the source of the influence, and the agent is responsible—ontologically, rather than causally—for the vehicle's existence. The vehicle, known as a *force,* is

[3] To be fair to Newton, he probably did not adhere to all the principles that we now associate with Newtonian mechanics. In particular—and as the quote in the introduction indicates—he did not nurse a bias for the micro. And he probably would not have embraced the one-scale model of the universe that is the target of my book.

[4] Relativity repairs certain deficits of Newtonian physics by adding the constraint that influence cannot be transmitted instantaneously, but must cross the intervening distances between source and patient.

also represented by a vector variable. And forces bring about observable variations and covariations (sync, if you will) in quantities.

2. The patients upon which forces act are quantities in their own right, whose magnitudes change in proportion to the magnitudes of forces impressed upon them. The alterations in magnitude too are continuous over time, according to Newton's Second Law.
3. Not every continuous course of magnitudes for a given quantity, beginning with the initial magnitude and ending with the final one, receives attention in a Newtonian analysis, where by contrast these alternative courses of influence are much the object of scrutiny in the second venerable branch of physical mechanics.

Analytical Mechanics

Analytical mechanics originated with the ideas of Gottfried Leibniz, Leonhard Euler, and Joseph-Louis Lagrange and culminated in the Hamilton-Jacobi equations, which are frequently (but quite illegitimately) treated as either reformulations or culminations of Newton's laws. Analytical mechanics conceives of alterations in time as a kind of wave disturbance. But these wave disturbances are not disturbances in a real medium with a three-dimensional spatial extension. They are, instead, disturbances in phase space—a space in which quantities (for example, momenta and positions) are represented in relation to one another. There are no counterparts to the roles of agent, vehicle, and patient in this formulation of mechanics. There are simply quantities, which undertake alterations together, and in keeping with very general principles that are not causal in nature. Distinguishing marks of analytical mechanics are as follows:

1. Conceptually, there are no agents of alteration. There are, instead, *variational principles*. Famous examples are Hamilton's principle (the integral over a system's path of the difference between kinetic and potential energies is always an extremum—either a maximum or a minimum—when compared with that same integral over all other paths) and Huygens's principle for optics that leads to Fermat's "principle of quickest arrival" (if light travels from one point to another point along the path of a light ray, it will arrive in the shortest possible amount of time).
2. Alternative courses of magnitudes for the quantities under study are of considerable concern. They are displayed in one of two ways: (1) by means of initial and final state configurations; or (2) by specifying "external" constraints upon allowable motion. These means of displaying the alternative courses of magnitude are typically called *boundary conditions*. Variational principles, in analytical mechanics, are applied to sets of these alternative

courses of magnitude, to select as distinguished (when there is a unique solution) an actual one.

As a matter of practicality, analytical treatments are routinely available even when the mechanical problem we want handled will not yield easily to an analysis of the forces purported to be in play. Alteration in time, on the analytical formulation, is explained without appeal to influence and its communication from one event or state of affairs to another.

The Newtonian treatment preserves the one-scale hypothesis by enlarging the supervenience basis—that is, by broadening the base of entities—and properties upon which macro depends, expanding the base to include such things as forces and their retinues. On an analytical treatment, by contrast, the actual trajectory emerges, according to Hamilton's Principle, as a result of there being a unique potential trajectory meeting all boundary conditions, and which possesses an *extremum* (typically a minimum) in the difference between potential and kinetic energies. Without prior specification of the boundary conditions, not only is the system to be treated underdescribed, but also the mechanical problem itself is *underspecified*. This is because the variational principles of analytical mechanics are in their essence *contrastive*: they select the unique trajectory that possesses an extremum of a particular characteristic, while presupposing that the set of all possible trajectories has an independent specification. This is why certain macroconditions and constraints on motion must be known prior to computation of the system's path through phase space, for not only do the macroconditions and constraints help to *identify* the contrast class of potential trajectories, they also *define* that set. And in analytical mechanics they are treated as independent, both of actual magnitudes of microquantities *and* of the principles of contrast that constitute the dynamical laws. (They are usually ascertained by macroscopic inspection or imposed through macromanipulation.) According to analytical mechanics, therefore, macroscopic structures—whether imposed externally or simply known to hold—are treated as limitations on the total number of dynamical micropossibilities, from "above," as it were.

Thus analytical mechanics refrains from introducing quantities, such as forces, that mediate between the actual and the possible—things that (in Newtonian mechanics) just barely manage to maintain the allegedly one-way dependence relation of macro upon micro. But in so doing, analytical mechanics allows violations of supervenience. It leaves gaps and no corresponding strictures for how they must be filled in. Indeed it subsequently proceeds to allow filling of these gaps by admitting macroquantities in the corps of degrees of freedom (a concept we shall discuss in time at some length and depth). In this way it also violates the strictures against the independence of macro (in ways that Newton the man might have embraced), allowing macroquantities to serve as degrees of freedom. How does this filling-in happen? The Hamiltonian approach is open to acknowledging—indeed must in routine applications—that certain relations among microquantities (macro quantities or boundary conditions) can precede in ontology the magnitudes of the micro relata. For analytical mechanics regards boundary conditions as (1) independent of dynamical law and (2) prior to it in ontology since analytical dynamics must treat boundary conditions as prior in existence to the operation of variational principles.

Our discussion here exposes the question of the nature of the priority and independence of boundary conditions in analytical mechanics. While we can't in this book give the question the attention it deserves, because we are not poised to contextualize the problem adequately, we must nonetheless acknowledge its philosophical urgency: in what does priority and independence of boundary conditions consist? And, more urgently, what could their priority mean? We must emphasize the point that analytical mechanics takes the priority and independence of boundary conditions as fundamental, not derived from other assumptions. The analytical conception of mechanics guarantees selection of a unique trajectory among those not initially prohibited. But the nature of the prohibitions (against which, as background, variational principles operate) must be construed as independent and prior to the theory itself. And consequently, macroquantities are construed as degrees of freedom, so that fewer microquantities are required to serve in such a capacity.[5] (In subsequent chapters, we will say that micro degrees of freedom are "sloughed off.") The metaphysical framework of analytical mechanics is thus clearly different from that of the more familiar Newtonian framework.

In the Newtonian framework, only actual magnitudes deserve to be called "properties." And magnitudes of both forces and macroquantities end up being calculated from them (if they're calculated at all) through some form of universal law pertaining to substances or the laws of dynamics they are bound to obey.[6] Merely potential magnitudes of microquantities do not inhere in anything. It is

[5] We can put the difference between the two modes of system portrayal another way. We can say that the Newtonian mode prefers to store information locally, where this is possible. Where it is not, it stores information in a law (like Newton's Second) that interrelates local information. By contrast, the analytical mode of portraying systems has no preferred means of storing information; it is just as happy to store information in an unlocalized format as it is to store it in a local one. In general, it prefers economy to format. If information is more economically stored nonlocally, then the nonlocal method of storage will be chosen.

[6] Do not be deceived: the name of "potential energy" in Newtonian physics attaches to a form of *actual* energy; the name simply designates its difference from energy that is realized by motion. Potential energy is not construed as a *form* of potential motion, in spite of being thought of as a *source* of potential motion. It is not a potential form of anything, but simply a form of energy independent of motion.

strange, and an expression of the pioneer's peculiar stern-heartedness toward the merely potential, to proclaim that *actual* magnitudes are properties of an actual body, but that *potential* magnitudes, if they are deserving of the name of properties at all, are second-rate or derivative properties—potential magnitudes are "properties-in-law" (pun very much intended) rather than properties in the proper sense of the term. Potential magnitudes are, according to this doctrine, properties in an attenuated sense only. In the analytical treatment, by contrast, actual and potential properties are both truly properties of systems. Potential trajectories are determined by a combination of constraints, macromanipulations, and confinements, which themselves do not receive treatment via dynamical laws.

Analytical mechanics' proposal to lump boundary conditions together with various macro features of a situated system (by conferring on both categories of information the same function within the theory—namely, the articulation of global possibilities) suggests that the analytical conception of system is "contextual." By this I mean it is inappropriate to request characterization of a physical system taken "in isolation," if by that we mean "without specification of boundary conditions." According to the conception of a system that emerges here, there is no such entity as *this system in splendid isolation* if this system is not already in splendid isolation.[7] In fact, there may be no such thing as a system in splendid isolation, period. Portrayal of a system, on the analytical approach, is always of a system situated, rather than of a system-in-itself.

Morals

Should the conception of an object include, not only its actual characters, but also its potential traits? For instance, I have brown eyes. I might have had eyes of a different color, or at any rate, so let's assume. Is that potential trait a character of *mine*, however merely potential? If so, how does this potential trait interact with the one I've actually got? These are questions that are very important in interpreting the states attributed by quantum theory to ordinary physical objects that we have known about for a long time. This issue is really at the heart of the difference between the quantum conception of objects and the classical conception.[8]

[7] Lanczos (1949, 4) puts it this way: "The analytical approach to the problem of motion is quite different [from the Newtonian approach]. The particle is no longer an isolated unit but part of a 'system.' A 'mechanical system' signifies an assembly of particles which interact with each other. The single particle has no significance; it is the system as a whole which counts."

[8] I have argued (2006) that certain treatments in statistical mechanics (at least in the hands of Wilfred Gibbs) utilized both.

3. Quantum Mechanics

The achievements of analytical mechanics exerted enormous influence upon the development of quantum mechanics. The cornerstone of quantum mechanics is the Schrödinger equation—it is the dynamics of the theory that ended up being key to unlocking atomic structure. The Schrödinger equation is a straightforward analogue of Hamilton's equation of motion in analytical mechanics. Erwin Schrödinger's principal innovation was to introduce a plane wave as a fundamental way of describing a certain system (namely, an electron) and then to generalize its use in a number of different ways, not least of which is that he insisted upon utilizing it as a representation of matter (namely, electrons) as well as light and energy waves. To justify the move, he borrowed heavily from analogical reasoning about waves as if they were particles and particles as if they were waves that he found in Max Planck, Einstein, and Louis de Broglie.

Schrödinger embarked on his own journey of discovery by searching initially for a proper wave equation to describe the electron. He sought something that would contain all the information about it that it was possible to have. And he noticed that wave functions (for instance, functions that describe the behavior of liquid in a wave tank) enjoy precisely that quality. A wave equation may not describe the motion of molecules in the tank medium, but it contains all the information about the *wave itself*—which is the disturbance cascade in the medium, construed as a kind of continuum. In the formalism that Schrödinger ultimately wrought—according to the standard interpretation—the quantum state, also called a wave function or state vector, is the most complete description that can be given of a physical system. And physicists typically entertain solutions to Schrödinger's equations that describe not only molecular, atomic, and subatomic systems, but also macroscopic systems, even unto the entire universe.

Scientists devise interpretations of their formal theories at the same time as they write down the theories themselves—on the fly. The innovators themselves don't always inscribe for us interpretations of the formalisms they innovate. (And sometimes what they do have to say by way of interpreting the steps in their reasoning is deficient—at least when taken all together as a package.) The task of careful, scrupulous interpretation is quite often taken up by people downstream of the originator, if the formalism proves useful. Schrödinger's did. That's why we care so much about what it means, why Schrödinger's initial interpretation of it has been scrutinized carefully—but ultimately found wanting. It was Max Born who came up with the probabilistic interpretation of the wave function that is second nature to physicists today. Born proposed that if you squared the wave function and normalized it (which means you set the total area under its curve to 1), you would essentially have the probability distribution of the particle. For example, if you consider just a single dimension of space (for simplicity), and consider a particle moving in that dimension, the chances of finding the particle between any two points along that dimension is given by a mathematical operation on the wave function for that interval (specifically, it is the integral of the absolute value of the square of the wave function over this interval). Born's interpretation has been widely accepted by the physics community (although it was never accepted by Schrödinger himself). Simply stated, the wave function speaks of possibilities and their strengths of being realized. Unlike a Newtonian model, it does not speak purely of actualities.

With that interpretation of the state function, quantum theory supplied us in chapter 1 with predictions and characterization of sync in the world. And the sync with which it supplies us goes (as we showed in chapter 1) counter to the proposal of universal supervenience of even the slightly-more-macro upon micro. We generated in the last chapter a very clean and simple proof of the inadequacy of a general principle of supervenience of slightly-more-macro upon micro.

Interpretation questions in quantum theory are still much discussed. But I'm less concerned here about interpretations of wave functions, or even about getting the principles underlying atomic structure exactly right, than I am about the general framework for conceptualizing how objects and their "local" traits are to be regarded vis-à-vis their environments. In particular, I'm concerned with the subject of independence, and whether it ever really makes sense to think about a completely independent entity—whose certain properties belong to it and it alone (belong locally, as it were), owing nothing whatever to the situation, the context, or the setting in which the object finds itself. We don't encounter truly independent entities in the social world: for instance, it can make quite good sense to say that human beings enjoy qualities such as kindness, good humor, or courage, but it is downright unintelligible to ask whether someone could be courageous (or good-humored, or whatever) outside the context of a human social environment, because outside such an environment, there are no "landmarks" or "backdrop" (some philosophers might like to use the word "norm" or "standard") against which some piece of behavior qualifies its performer as either courageous or not. It's not that the behaviors in question would be unavailable; it's just that their qualifications, as courageous or kind or whatever, would be ineligible if the norms for them did not exist somewhere. Similarly, human beings speak languages, but no one speaks a language without being trained in it. The question is just barely intelligible whether the word "potato" would mean what it does in my mouth, if I hadn't learned English from an English-speaking community. (And to the extent that the question even makes sense, we must answer it with a "no.") This principle can easily be extended to biological contexts generally: no member of a real-life population of animals can be a fully functioning member of that population unless the animal is appropriately conditioned—and sometimes that means quite literally participating fully in the evolutionary history of that population.

Why not take this line of reasoning to its extreme: nothing exists as the thing that it is—not pigs, not dinosaurs, not even electrons—except as it is appropriately so conditioned by an appropriate structure. There are no independent entities. This theme is consonant with the analytical mechanics framework that we discussed earlier and saw mirrored in quantum mechanics. It makes excellently good sense of Cooper pairing—of the bosonic character of a pair of fermions, in the context of a sea of oppositely charged particles, which we encountered in the explanation of superconduction in the last chapter. We will now embark on a couple of further explorations that emphasize and support this theme.

4. Approximation

First, a bit more on the structure of matter and how quantum theory manages to make progress on illuminating it. A familiar planetary model of the atom is taught in preuniversity educational settings almost everywhere. This is the Bohr model. It comes with a number of ad hoc assumptions (to make it consistent with atomic spectral emission data that was available to Bohr when he delivered himself of this model). So it's a good model, but lacking in certain theoretical virtues.

Along comes Schrödinger, who tells us we can just assume that the whole electron-bound-to-a-nucleus business is just like an oscillator—like a mass on a spring or water in a wave tank: a generator of waves. In his way of thinking about it, it's really the waves we are interested in describing—the waves are the reality of the whole business. And so we can apply an analogue of Hamiltonian mechanics to determine what the (time-stable) orbits of electrons look like. Solve for the "stationary" states—the ones that don't change over time—and that's atomic structure! We thus assume that the solution to the analogue of an oscillator equation we write down describes the characters of a particle—the electron. We assume that the solution to that wave equation contains all the information about that electron that there is to be had. (In fact, when we assume differently we generate false predictions.)

This is nice example, but it is also an easy one. One electron, one nucleus—add them up and you have just two bodies, an easy enough dynamical problem to solve. But solving any kind of equation of motion for a system of more than two bodies acting on one another is unbelievably hard—there are too many variables to keep track of and too many cross-effects. On top of there being too many equations, the equations we end up with are nonlinear, and that means they are just too hard to solve (nonlinear equations are technical buggers in real life).

So when you get more than a couple of bodies in a system, you have to do something besides solve basic equations of motion involving as-if oscillators. One thing that physicists are always looking for is an approximation or approximation method. Now an approximation is not just a guess, it's a guess with a principle behind it. It has to be the kind of guess that makes the problem easier to solve. Also, it can't be so close to the answer itself that it leaves no "gap" between the guess and the true answer. In other words, the guess has to allow the journey-via-approximation to illuminate the answer. Approximation itself, when it succeeds, is nothing short of deep insight into what is of real significance in a problem—what is at the heart of the process being illuminated.

Now, suppose that we can make other problems look like Schrödinger's two-body problem, even if they weren't really composed of just two bodies. Then we might be able to solve them too. This is the idea behind the so-called *mean-field approximation.*

What can we do to approximate the interactions of many bodies, to illuminate the "heart" of this kind of situation, but nonetheless simplify it so that in form the equation is like that of a two-body problem? One thing to do is to simplify the formulas for the many interactions, without doing anything to avoid having to take into account the number of interactions. That is the style of modeling we put on display in chapter 1 with the linked model of flocking. Linked modeling is a good framework for that style of approximation: it keeps the array on display and simplifies the nature of the linkages. But another approach to approximation is to try to lower the number of interactants themselves.

There are many approximations of this sort that one can try. Some work better than others for different purposes. I want to discuss a couple of these in order to make it possible to raise an important philosophical question.

5. Mean-Field Approximation

Suppose you're trying to model the behavior of a certain body, which is embedded in a sea of other bodies. That sea may be more like a pudding than a rigid crystalline array, or it may be much more diffuse than either one, more like a liquid in a glass or a swirl in a gaseous miasm. So you should choose your approximation accordingly. One conception of this situation of many-on-one is that the body you're interested in will interact with every other body in the system—with every molecule or atom or what have you in it. That conception may give way to another one that I will try to characterize by first articulating one form of it.

The main conception of the mean-field approximation (also known as *Hartree-Foch theory*), which originated very shortly after the fundamental principles of quantum theory itself, was born around 1925. The idea is to replace all influences upon a certain body (the body whose behavior you're interested in modeling with a net or effective interaction—a field that somehow summarizes the effects of all other agents in a point-by-point way. (This is why it is sometimes called a *molecular field.*) If you could achieve this result, you would reduce any multibody problem effectively to a one-body problem—much like the one Schrödinger solved for the electron of a hydrogen atom. (To see this, you have to recognize that his conception of the situation is that the one body—the massive nucleus—is fixed and is the source of the field.)

The Nobel physicist Philip Anderson, in his 1963 classic *Concepts in Solids,* writes, "One may get the impression that modern many-body theory, of which one hears so much, goes far beyond Hartree-Fock, and that therefore we should not bother with such old-fashioned stuff. This is not at all true—in fact, modern many-body theory has mostly just served to show us how, where, and when to use Hartree-Fock theory and how flexible and useful a technique it can be" (1997, 9). In fact, there are techniques that are conceptually different from Hartree-Fock, and which were devised since Anderson's writing. The motivation for these more recent techniques was to explain certain aspects of the large-scale behavior of matter at critical temperatures and pressures—namely that certain substances that differ vastly in their molecular structures, constitutions, and organizations can nonetheless share entire portfolios of basic thermodynamic characteristics. (We will have occasion to tangle with these methods, known as renormalization methods, in chapter 8.) But the existence of these techniques does not detract from Anderson's point. The Hartree-Fock approximation does something conceptually very important and allows us to view a beautiful intermediate model between the Newtonian model and the analytical model.

Auyang (1998) describes the Hartree-Fock model of an electron in a sea of positive ions this way: "the only influence an electron has on the other electrons is through its contribution to the effective potential. The effective potential felt by [the] electron depends only on its own coordinates, for the contributions of all other electrons are averaged over. Since the same consideration applies to all electrons, the effective potential serves as a common situation whose effect varies systematically according to the positions of individual electrons" (Auyang 1998, 129). I'm quoting Auyang because she is very careful with her usage of the language of "common situation" and because she holds a very interesting position on the nature of approximations that will become valuable for us to consider later.

In the body-within-a-sea-of-bodies model (à la the Newtonian schema of the problem, the first approximation one typically thinks of when considering the problem), it might make sense to construe the entities in the model as more or less independent. But in a mean-field approximation, it no longer makes sense to construe entities this way. The behaviors and features of interacting bodies—assuming that there are such features and behaviors—are now globalized and contextualized. They are quite literally made attributes of the spatial expanse of interest. So there is some intermingling of entity traits with traits of context (with the "boundary and initial conditions" that are sometimes treated as one word in physics texts). Figure and ground blur and merge. This is the hallmark of wave/particle duality. Still, so long as the situation doesn't change significantly—for example, so long as no new entities enter the scene—we understand the nature of our model; we know where independent body (the target of analysis) ends and context begins.

But now consider a wrinkle: a new body shows up on the scene and threatens to impinge on the situation. How do we conceptualize this process? How is this impinging phenomenon to be conceptualized vis-à-vis the body that is the target of analysis? What about its impact on the mean field? We have options. I'll name a few. (1) We can treat the impinging entity as a random fluctuation (ignore it, or use statistical analysis to handle it and its like). (2) We can figure out a way to model its effects as dynamical effects on the field (absorb it). Or (3) we can treat it as a figure against a ground—modeling it explicitly as another particle (which can be very tricky, especially if it's the first of many such impingements). And so, as we keep making modeling decisions of this kind, perhaps even on an ad hoc basis, figure and ground blur further together. Where is the context, where are the entities? At the extreme of this process, entities disappear completely, and the model is entirely of wave phenomena. For instance, when we consider sound, the model contains nothing at all that resembles a body.

Sound waves are waves purely. Conceptually speaking, waves are simply disturbances, literally disembodied as they move through space, carrying energy from one point to another. In the case of a wave through a material medium, this is often with no (permanent) displacement of the medium. An initial displacement starts the process. And this creates a cascade of disturbances (oscillations or vibrations). A *transverse* wave consists of disturbances (oscillations) at right angles to the direction of energy (wave) motion, and a *longitudinal* wave (like a sound wave) consists of vibrations in the same direction as the wave motion. But the wave itself is independent of all the particulates that are oscillating or vibrating; it can move at a speed that's not related to the motions of the particles that are in motion; and all its properties are once again independent of the properties of the particles. It makes no conceptual sense to speak of waves being constituted by matter. The categories of wave and matter are distinct. They are *duals* of each other—something akin to conceptual opposites. And yet treating the one framework of analysis as analogous to the other has been extremely productive in the history of the field of physics.

Obviously the motion of particles has something to do with the properties of the waves that are somehow put into place via the particles' activities. But how constrained the properties of the waves are by the properties of the medium in which they are traveling is contingent and varies according to circumstance. Sometimes the substantive constraints are imposed, not by the medium itself (though of course this does contribute to the quality of the wave), but by the boundary conditions of the medium. When this is the case, the characteristics of the wave (musical tones, for instance) can be changed very easily by changing the boundary conditions (changing the length of a vibrating string or column of air, for instance). You can easily make a musical instrument of the phenomenon. So if you think of the material and the wave features as "levels," you end up having to embrace the principle that you can't always say what the wave properties will be, even if you know what the material is that it's made of. That is because even if the material basis imposes substantial constraints on the kinds of waves you can get (and obviously it must), it might well be that the "boundary and initial conditions" swamp everything—that they are much more important to understanding the wave characters than the material basis is.

6. The Meaning of Approximation

The electron bound in a hydrogen atom (the situated entity) is quite a different thing from the free electron. In quantum mechanics they are even associated with different quantum numbers. So it seems all right to talk about these as very different things. But what does that mean? And what does it mean for the proposition we are entertaining: that there might be no such thing as a genuinely independent entity?

Auyang (1998) warns us not to be confused, when we are using an independent-particle model like the mean-field approximation, into believing that there really are these independent things that react independently to external "mean fields." I wish to go one step further and ask: if the situated entity—the bound electron, say—is what it is because of the context, for instance since its energies and possibilities are delimited by its context, why isn't that true of all characters universally? After all, there are no truly free electrons. It might be that some come very close—possibly electrons on the edge of the universe, for example, if our universe has an edge—but never really achieve contextlessness. Why isn't context *always* a contributor to a thing's characters? Why aren't characters *always* conditioned by context?

My proposal is this: everything is qualified. In fact, that's the best way to understand a trait or character—as a qualification or restriction upon the thing "characterized," forced on it at least partly from the outside. This is not to say that the electron doesn't bring *something* to its situation; but this fact doesn't make the electron an independent thing. There are no independent things at all. And that's another reason to give up on the notion of supervenience: there is no local and independent supervenience basis, when all is said and done, because everything is qualified by everything else. And unsurprisingly, this is a very hard thing to model. So we have to pick the most important qualifiers of a thing (the most dominant effects, as they say), in order to get a reasonable treatment of it.

We've been talking about the fields we have characterized as approximations. A wave is a changing field. So perhaps the "wave" aspects of things are all approximations. They are the "cumulative" effects of bodies doing their thing à la the Newtonian picture of forces.

But what does it mean to say that some model is just an approximation? That things really are not exactly as the model says they are? That there really are no fields or waves? This would fly in the face of all of physics, especially so-called particle physics (a misnomer since everything in particle physics is a manifestation of a quantum field). Perhaps one had better ask: how is the notion of an approximation different from maintaining, instead, that there is a duality—that something is a bit like this and a bit like that too? These questions are too big to be settled here. But it bears reflecting that there is no way for us, as the gargantuan beasts we are, to perform any feat on an independent particle, if one existed, *directly*. All our actions will always be actions upon a mean field—they will be actions upon a macro trait or one of the many "boundary or initial conditions." It might be true that as I swing my arm through its natural walking arc, I displace an individual particle. But this action of mine (assuming it is an action) is indirect: it is because I act "first" (in the ontological rather than temporal sense) on the population of particles (many tens of millions), of which that one individual particle is a member, that I act "subsequently" upon it to displace it from an original position with respect to me.

In the context of this observation, we are reminded of the difference between the analytical mechanics formulation (with macro boundary and initial conditions as independent degrees of freedom) and the Newtonian formulation (in which no macro can be independent). They are different, just as I am saying that action at the macro scale is different from action on a micro scale. They are not equivalent. The former proclaims that macro features are active, one upon the other, but also and more indirectly upon micro features. The latter acknowledges nothing but activity amongst the micro.

I imagine Auyang objecting at this point. Don't be confused, she might admonish, by the fact that we can approximate interactions with fewer less-interacting particles, and conclude that the complexity doesn't exist. Consider that we can do this sort of thing in a wave framework: we can apply a wave framework and then translate our findings into "particles." For example, imagine (just for simplicity) that a wave tank full of fluid, with its characteristic "boundary and initial conditions," can sustain two modes of standing waves (like two frequencies in a vibrating column of air). These two "solutions" to the system's relevant differential equations will correspond to two different quantities of energy that one imagines can get pumped into the system in a stable way. By analogy with the hydrogen model, we can think of these as different "shells" that a particle (let's call it a "phonon") can occupy. The phonon is a real concept in physics, where it's sometimes also called a "quasiparticle." A phonon is an excited state in the quantum mechanical quantization of the modes of vibrations of elastic structures of interacting particles. Consequently, the concept plays a major role in many of the physical properties of solids. (It's a special type of vibrational motion, in which a lattice uniformly oscillates at a certain frequency, which in classical physics is called a "normal" mode. It is useful to be able to find a normal mode, because sums of normal modes are also solutions to the differential equation—they obey a superposition principle. The modes are *normal* in the sense that they are independent: an excitation of one mode will never arise from or result in an excitation of a different mode.)

So the point of our imaginary objector (Auyang) is this: we should not be confused by the formalism into thinking that there really are phonons. They are "as-ifs"; they're a *façon de parler*. There are in reality no such things as phonons.

No? And if we deny their existence, don't we also have to deny the existence of collective excitations? Aren't they equally as-if? Which is right: that there are collective excitations, or that there are phonons? And what do we gain when we affirm one but not the other? Which is the "true" article? And how does this answer help us advance our understanding of things? These are not easy questions. Confusion is bound to happen—and usually we will be tricked into choosing the wrong option.

This is all the more reason to be careful about how we regard the different formulations of mechanics. They do say very different things about degrees of freedom in the universe. And we cannot simply move between them indifferently, as if they amounted to the same thing. In fact no one can really maintain that the pictures are equally correct. The Newtonian framework is strictly speaking incorrect; and if the quantum mechanical framework is incorrect too, it's at any rate the best we've got so far. Quantum mechanics is fundamentally an analytical dynamics framework—contrasting with the Newtonian mechanics as we described above. Quantum mechanics, as we discussed in the last chapter, is a synced model.[9] If quantum mechanics is indispensable, then we cannot do without the wave dimension of the dual, against which the particle (which is the fundament of Newtonian modeling) is the opposite pole. And certainly we seem unable to do without the particulate model as well. What does this mean for the concept of an independent entity?

[9] It is clearly *not* a linked model. One can often go from linked to synced—but the reverse is never assured. That is one reading of Bell's Theorem.

I think it means that there are no truly independent entities, but that sometimes it makes sense to model things as if there were. So it is not the *wave* aspect of a phenomenon that is the approximation; if anything, it is the *independent entity* aspect that is the approximation. One can get misdirected by the mean-field method into thinking that it's the "sea of molecules" that's being approximated by some other entity, when in reality it is the anticharged body that is the thing that's doing the approximating. In truth, everything is more like the field than it is like the particulate. I do not claim to have offered decisive evidence for this position. I simply claim to have shown that this position is no more tenuous than the converse. It is the opposition to the one-scale model of the universe—which insists that the smallest bits of the universe are truly independent, and indeed the only ones that have "fundamental" reality. For the sake of this nameless position (which I might as well refer to also as a scale-free model) I am tenuously repelling the more familiar one.

This point makes it easier to understand how there can be activity at all scales of the universe. There are wave phenomena at all scales of the universe. There is absolutely no reason to suspect that one scale of wave phenomena is in any way more privileged, in the sense of being more "active" or "influential," than any other. This explains why supervenience is always contingent and frequently false.

7. Making Stuff and Doing Things: Resheathing

In what regards could it matter to us whether or not supervenience holds? It would matter if we had to make things by cobbling other (tinier) things together—if we had to build things up one atom or molecule at a time, as it were. But that's *not* how we make stuff or do things. Making things, for us (in all but the most unusual cases), requires a macro process. You can't make a magnet by putting dipoles together "one at a time." Cobbling is simply not available. You have to act on them all at once. That's what resheathing is: acting at a higher scale so as to *entrain* a new state upon things at a lower scale. To do that, one manipulates macro states of various collectives. One changes their "boundary and initial conditions." This idea is paramount. Resheathing is different from cobbling, metaphysically speaking. It often involves the use of "context" to impose order that issues in "waves" downstream of the original action. This is what I now hope to prove to the reader, via an exercise in magnet-making.

Never mind that magnetism is a quantum phenomenon (which it is). Let's consider what it takes to make a magnet. One needs to start with a material that is appropriately susceptible to magnetization—a ferromagnetic substance. Ferromagnetism is the phenomenon by which materials, such as iron, when placed in an external magnetic field, become magnetized and remain magnetized for a period after the material is no longer in the field. A substance that is susceptible to being magnetized can be transformed into a magnet as follows. The process relies upon a property of magnetic materials known as *hysteresis* (the lagging of an effect behind its cause). When an external magnetic field is applied to a ferromagnetic substance, the item "absorbs" some of the external field: when the external field is removed, the magnet will maintain some of the structure that was imposed on it via the field—it becomes magnetized. The substance changes magnetic phase. This change in phase of a ferromagnetic material is sometimes referred to as memory, because it resembles the learning process with which each of us is familiar as a subject or agent. Ferromagnets are like entities that learn, in that the changes wrought in them, via "teaching" episodes, outlast the duration of the training process. In other words, teaching episodes result in the entity's retaining (some of) the entrained conditions even after the teaching device is removed and the external pretraining conditions are restored. This means that the phase condition is not strictly linearly dependent upon the external "standing" conditions, but depends also upon the substance's history.

What the history does is put the susceptible substance into one of its eligible material phases. To do this in real time we simply take a ferromagnetic substance through the "training regime" described above. Keep in mind, we don't take the susceptible substance one unit—one atom—at a time. We don't cobble together anything at the purported "scale of activity of the universe." And the reasons are clear, as we can explain directly.

Herbert Simon, as we will have occasion to discuss again in chapter 3, believed that certain features of certain systems could not be captured correctly by analysis of their parts taken together with the concurrent interactions among them.[10] To build a complex system (as nature routinely does), one (like nature) proceeds in stages, with the result that at the end of each stage what is constructed must possess a stable structure (so as to "hold still" while the next phase of operations is launched). Without these intervals or layers of stability, complexity (according to Simon) is unsustainable.

To build a complex system in real time, one typically has to put together the right parts, at the right time, in the right sequence of operations. And so an atom-by-atom algorithm (however conceivable in the abstract—in principle, as philosophers say) is not a recipe for making stuff, even if we're talking about a purely physical system.

The atom-by-atom focus is akin to thinking that a painting consists of a pattern of colors. If we can duplicate the color pattern, we've duplicated all the "natural" properties of the painting, except for its history. If we could, for instance,

[10] See Simon (1973). I present a treatment of this in Thalos (2007), from which I draw in chapter 3.

automate a means of analyzing colors on a canvas, down to as fine a grain as we like, and subsequently devise and automate a way of conveying color to a new canvas according to this pattern, we will have duplicated the painting. But I can assure you the this much is true: if the original is a watercolor, this process will not work unless care is taken to ensure that the process of laying down watercolor is sequenced properly. And the proper sequencing just might require the painter and her brush! This can't be done atom by atom.

This shows why it doesn't matter to us whether larger-scale features of the world supervene on smaller-scale features. We don't make the larger scale by cobbling together things at the smaller scale. It should matter much more to us if we actually *couldn't* make things by acting on macro on the larger scale. If that were the case, then cobbling would be the only way to make things. That would be the true test of supervenience: test whether there is only one way to make things at a large scale. It's a shame no one has proposed this test. But of course there is a good reason why no one has proposed it—for there is always a certain as-if response. The response is that *if* one could act at a macro level, it would be *via* bringing about the micro basis of the macro condition desired. This response is explicit in Kim (1999, 24): "To cause any property (except those at the very bottom level) to be instantiated, you must cause the basal conditions from which it arises (either as an emergent or as a resultant)." (And this, as Kim notes, is a formulation of downward causation.) But this is sheer prejudice against the reality of the macro: why isn't it true that one brings about a macro condition directly, or that one brings about a micro condition by bringing about a macro condition first? The fundamental problem is this: both statements, when interpreted in the ways that seem familiar, involve employing supervenientist metaphysics, which is precisely the claim at issue. Even the language of bringing about a macro condition "directly" suggests one could have done it also "indirectly"—via a route that passed "first" through modifying micro conditions. And any proposed success of a test of whether we can act at a macro level to do things and make stuff would be interpreted as a failure to prove the intended point. This is why we are taking an alternative pathway with this difficult question. We should like to teach ourselves to interpret activity at macro scales in more ways than one.

8. Resheathing Is Everywhere

We talk in order to communicate. And the communication (the message) can be an instrument for achieving something else—a manipulation of some further kind. For instance, I can tell you that I shall be home tonight, as a signal that I would welcome your visit. It might seem as though I achieve this simply by making sounds in my throat that manage somehow to reach your ears. That would be an oversimplification: for I accomplish may aim by constructing an entire context whereby you know how to make sense of the sounds. If you speak only French, for instance, I'd be wasting my time articulating any message in Mandarin. And if you don't know where I live, I'd be better off sending a message with that further information. And if you're slightly autistic, I'd best spell out exactly what I'm hoping for, in case you can't tote up certain sums of two and two.

In constructing contexts, I act at a very high scale. And to do that I rely upon the entities in my context having been "trained up" in certain ways. I actively recruit or invoke their training. If you are untrained in the conventions of my culture, then when I say that I shall be home tonight you might get a quite different message than would be received by someone with exactly your current motivations, but also familiar with my culture. I have to pick the right context in which to communicate a message that your particular ears can interpret in the way I want you to interpret it.

Sometimes a person seeks to achieve a result by breaking rules or conventions, such as when a normally polite person produces a profanity, or when a player on the stage addresses the audience rather than other players or props. Breaks from convention also send messages, though messages of a much more ambiguous nature than messages sent using channels made exactly for such messages. Artistry is involved in choosing the right medium for a message: some ways of sending messages are more artistic than others, because they are more powerful. This doesn't mean that the alternate ways of communication—the rule-breakings themselves—aren't also governed by their own rules. If there were no rules whatsoever, we would have a hard time saying there was a message at all.

It's the same at much smaller scales of transaction. One can create an electrical current using the "common" pathway of ordinary conduction (though metals). That requires picking the right kind of physical context (metal wires, etc). Or one can create electrical current using a different pathway, via superconduction, by changing the context (lowering temperatures, using different materials that ordinarily do not even conduct at all, since good conductors are very poor superconductors). And one has to do it by breaking with ordinary conventions of electrical conduction—to create a context that calls on principles of quantum mechanics. All of this is done by creating an alternative context at a very high scale: I might lower temperatures, make wires of an appropriate materials, and so on. But I don't do anything at a micro level. True: when I try to explain how my method works, I need to talk about things that are going on at the micro level. (Auyang 1998 refers to this as *synthetic microanalysis*—a term I like very much.) This is decidedly not to say that all the real action happens at the micro "level." It might be understandable how someone would be confused by the fact that I invoke processes involving the micro to explain superconduction—it's understandable that this someone might suppose that ultimately that's where everything happens. But we have said enough by this point about this confusion. It's time now to set the record straight.

In summary, it looks as though the fabrication of something that didn't exist before, even if the thing is a micro physical system, requires at the same time bringing about both micro and (some) macro aspects. And it is no different whether the subject is magnetism or speech communication. So, what is there to divide the micro as fundamental from the rest as in some way different, and thus problematic? There are no clean lines. The universe is active at every scale. I am using the terms "sheathing" and "resheathing" to refer to activities and processes whereby activity at some scale initiates a learning process, at the conclusion of which a new condition endures (however briefly), even though the teaching activity might be removed.

Perhaps somewhat surprisingly, destroying a sheath (unlike destroying a cobble) typically also requires a macro process. Or perhaps this is not so surprising, since, as Simon insisted, there are inherent "stabilizers" in the macro process that allow us to build things in the first place.

If we take seriously the proposition that situated entities (which, if I'm right, are all the entities that there are) are not independent entities in any traditional sense, and with it the philosophical proposition that most if not all characters are nonintrinsic, then we will be obliged to think about objects, and the entire world, rather differently. In this new world, things come into existence not only via the pathway of components being cobbled together, but also via the pathway of larger-scale traits ("boundary and initial conditions") which impose new structure and order on entities within their boundaries, at a somewhat smaller scale, engulfing entire populations of sheep in wolves' clothing (as it were).

It's fortunate that new things can come onto the scene this way, because we humans would otherwise have no way to make things happen or to make new objects. A theory of makings of this sort requires that we take seriously the actions of globalized characters upon more localized characters—a downward causation, if you will. We have no general theories of activities of this sort. And if philosophers do not take up the project, there is no one else who can. But sadly most philosophers are averse to the scale-free universe in which these things can happen. This book will help to lay down foundations for accounts of this sort by articulating a theory of science that helps to make the analysis possible. We will shift gears now to talk about the foundations of science.

PART TWO

PHILOSOPHICAL FOUNDATIONS OF SCIENCE AND METAPHYSICS

3

Multiple Conceptions of Fundamentality

> I assume . . . that matter is ultimately particulate. I assume that every material thing is composed of things that have no proper parts: "elementary particles" or "mereological atoms" or "metaphysical simples."
>
> —Peter van Inwagen, *Material Beings* (1990)

It is customary in philosophy nowadays to make the sorts of assumptions Peter van Inwagen explicitly articulates in the epigraph to this chapter. Making such assumptions about what is fundamental seems to mark one as especially rigorous, serious, or in some other way no-nonsense. But while making explicit one's assumptions is admirable, examining them is even more philosophically important. Critical examination of these assumptions is rare. This chapter will be dedicated to examining the class of assumptions made on the topic of fundamental ontology, and how these assumptions affect the grounding of scientific inquiry. The moral of the examination is that van Inwagen's assumptions, while common, are neither especially sound nor especially helpful as a theory of what is fundamental.

1. Foundationalist Instincts

The ancient atomists Leucippus and Democritus theorized that there are two fundamental and opposite constituents of the natural world: indivisible bodies—atoms—and void. Void, they held, is simply the absence or negation of body. Atoms are by their nature intrinsically unchangeable; they can only move about in the void and combine into different clusters. Since atoms are separated by void, they cannot fuse, but must by necessity rebound upon impact—leaving one another unchanged. (Collisions are *elastic*, as the physicists say.)

Aristotle rejected Democritean atomism, on the grounds that his theory of substance (and especially the internal potentialities of substance) is fundamentally inconsistent with unchangeable atoms. Many centuries later Descartes proposed a "corpuscular" theory of matter that bears only a superficial resemblance to Democritean atomism. In Democritean atomism the atom is indivisible, whereas Cartesian atomism holds that, since every spatial extension is divisible in thought, and God has every conceivable power, God has the power to divide corpuscles.

Modern atomic theory retains elements of both its predecessors: physicists have split the atom, but subsequently have discovered that atomic division results in substances fundamentally different from those with which one began, a finding with an Aristotelian cast to it.

One moral of this history is that our appreciation of matter only improves with understanding of the smaller components and their composition. Our theories become more nuanced, and our concepts more refined. But ultimately this process comes to an end. At least, this is the faith van Inwagen expresses. It is faith in a fundamental, bargain-basement reality, below which there is no more going down, and where we might finally stop the whirlwind descent.

But is that correct? What evidence is there that he is right? Perhaps van Inwagen feels that the evidence for this is sufficient. Jonathan Schaffer (2003, 505) disagrees:

> We do not, right now, have any evidence for atomism. We now have no evidence that there will be a final theory, no evidence that such a theory will postulate anything that could serve as a mereological atom, and no evidence that such a theoretical postulate will correspond to an ontological atom as opposed to a boringly decomposable composite. Evidence for fundamentality is lacking thrice over.

To the extent that there is pervasive confidence in a ground floor of entities, that confidence must be construed as rooted in simple faith.

My point for now is simply that there is a cast of mind that insists upon the bargain basement, for which anything else will come as a disappointment. Anything else will signal an ontological insufficiency—a shortfall in our scientific understanding.

Is "cast of mind" all there is to van Inwagen's faith? Is it merely inclination or can something further be said in its favor? And is there anything wrong with such an inclination? This question is ultimately less important than the suggestion in it—namely, that *scientific understanding* demands a bargain basement. This suggestion is that the basic parts of something explain all the higher-scale functions and processes to which it contributes. This is *ontological foundationalism*. Van Inwagen's inclination, if its status is also an imperative, is perhaps ultimately founded on this principle. It is this foundationalist principle that demands philosophical assessment. I shall refer to this principle as the *building-block conception of fundamentality*.

The building-block principle of fundamentality is interestingly related to another principle—the principle of reduction—to which we now turn.

2. Reductionist Instincts

The building-block conception of fundamentality derives from a conception of the unity of science with roots in conceptual analysis.[1] As noted in the introduction, this model of unity originated within the Vienna Circle and was core to Carnap's systematic philosophy. On this conception, the sciences are unified through the unity of their subject matters, which is evident in the relations between their concepts: the concepts of the sciences stand in hierarchy relations, corresponding to the hierarchy of the sciences themselves. Carnap's system was characterized by a concern for logical constructions out of basic concepts in axiomatic structures, and rigorous reductive logical connections between conceptual categories. This orientation led him to a privileging of a "most basic stuff" and a preference for physics as a privileged locus of such stuff. This consequently led him to a hierarchy of conceptual structures for the sciences.

Many construals have emerged since Carnap's time as to how precisely to make sense of the relations between the conceptual apparatuses of different sciences. The most prominent ones today revolve around the notions of supervenience and emergence, building implicitly upon an assumption that there are fundamental entities or *atoms*.

From here on I shall use the term *atomism* to refer to the doctrine that there is a ground floor of *entities*, or *building blocks*, from which all other entities are built. I will use the term "atom" in its original sense, as the smallest possible constituent, the smallest indivisible "part" that anything may possess. The fuss, when once the proposition of atomism is accepted, typically revolves around how to account for the "granularity" of the world—into such things as medium-sized objects and living organisms.[2]

The importance of atoms, in the economy of scientific thought, is precisely this: identifying the atoms helps us identify the place where we can stop asking why. Once we know what the atoms are and the rules that govern them, we will have a "final" answer to the question of why things happen as they do. For, obviously, once the "final" theory has been ascertained, it makes no further sense to

[1] Bealer (mss.) gives a penetrating treatment of the relationship between reduction and analysis, while the tension between the reductionistic and nonreductionistic ideals among philosophers concerned with the unity of science is brought out nicely in Cat (2007); cf. also Batterman (1995).

[2] Van Inwagen (1990) resists the "reality" of these layers, but he is an outlier.

ask, "But why do these atoms behave like this?" Roughly, then, atoms are "ultimate answers" to why questions. That's why they're "ultimate" and "fundamental."[3]

While insistence upon a theory of science as unified in Carnapian terms has waned, focus upon its legacy of *levels* has only intensified. The conception of levels provides a corresponding conception of fundamentality, without demanding the unity of science. Supervenience enthusiasts and emergentists are vying for a theory of the world in what may be termed an "accumulationist" or "sedimentationist" model—articulating sufficiency relations between a (putatively large) set of facts at a "fundamental" level and those nonfundamental levels that depend on them. This is essential, they feel, to preserving a scientific view of the world. The effort in recent decades has been focused upon the formulation of nonreductive physicalism in a climate of antireductionism, as we saw in chapter 1. Fundamentality in atomistic terms has been viewed as key to a respectable philosophy of the world.

But there is a second conception of fundamentality that one could adopt. This chapter will articulate this alternative conception.

3. A Second Conception of Fundamentality

A *fundamental theory* in the atomistic terms just sketched, assuming we had one and it were as true as it is devoutly wished-for, is not likely to shed much light on (for instance) the lives of biological entities. And so the building-block doctrine of fundamentality deals with ontology in the most narrow sense. It is not concerned with metaphysics more broadly, not (for instance) concerned with the nature of the differences in the layers of order, nor the scales in which they are manifested, nor with their theoretical reliance upon one another, nor finally with the various autonomies they enjoy. So obviously a fundamental theory in building-block terms, assuming we had one and it were true, is not a theory of science in the Vienna Circle's sense—whatever its merits as a theory of entities. Advocates of levels believe this is an advantage. For such a theory would be nowise concerned with the ideas, categories, or theories that are *key* to understanding a domain or phenomenon. *Key-hood* is (itself) the key to *epistemological* fundamentality.[4] So to the extent that a "theory of everything" (TOE), if there is such

[3] I'm not using the term "atom" in its contemporary sense as an object with a nucleus and bound electrons, but in its original sense as the smallest indivisible—and thus eternal—building block of matter.

[4] The standard way of conceiving the relationship is to say that "laws" framed using the key concepts in one (higher-scale) domain do not yield to efforts to "reduce" them to laws framed using key concepts in another (lower-scale) domain; thus, the key concepts of the higher-scale refer to causally autonomous features and are ontologically fundamental. I do not endorse this view of the relationship between epistemic and ontological fundamentality, but nothing I shall say here depends on taking either position on the matter.

a thing, will *not* shed light on the nature of, for example, biological phenomena, we shall require some analysis—presumably in philosophical terms—of the relationship between the entities inhabiting the scale of description of the TOE and those residing elsewhere. It will have to be a metaphysic.

The building-block conception of fundamentality I have just sketched is an eligible element of such a metaphysic, happily reliant upon the (ultimate) identification of smallest (physical) independents—the *simples*. It is thus a theory of *ontological simplicity as ontological fundamentality*. According to this theory, simplicity is counterposed to *complexity*, which is often characterized in terms of the interdependence of numerous factors or features upon one another.

By contrast, disciplines that draw on engineering concepts focus upon ineliminably interdependent aspects of the world—other features of ontology—in order to characterize and explain important realities operating at higher scales. This suggests—and I will indeed defend—the idea that theories reliant upon ineliminable interdependencies for expository purposes will have a very different conception of ontological fundamentality than the now-pervasive conception of fundamentality as ontological simplicity.

The assumption that there are absolute, non-context-relative simples—"atoms" that are basic independents—is at the heart of the reductionistic orientation we have been sketching. Specifically, atomism embraces the temptation to treat some one object type (or perhaps a small menagerie of types) as inalienably *independent*. This is a commitment that amounts to adoption of an *absolute* conception of independence, as well as an absolute conception of atomhood. Atomism is thus committed to the idea that independence belongs only to the atoms.

Along with this commitment comes another—a commitment to treating pervasive correlations or coordinations among the purportedly independent parties (the atoms) as a matter of incidental, "external," or "ecological" interactions between them, rather than as systemic and inalienable properties of confederations among them. Atomism thus renders relations between fundamentals *external*—rather than acceding to the possibility that relations may enter *internally* into the makeup of something that is itself, and despite being nonatomic, a fundamental.[5]

Thus assessment of atomism, to be thoroughgoing, must evaluate the entire reductionist package of model-building strategies, which includes (1) independent

[5] One might complain here that relations have been traditionally construed as supervening on the entities they relate, and hence that any time "parts" are in evidence, there are at least two orders or levels of entities. Hence the distinction between internal and external is problematic. The obvious reply to this complaint is that the conception of supervenience is itself the problem, and that leaving the nature of relations open to inquiry is the right way to proceed in the debate. We shall discuss the nature of relations in good time.

building blocks and (2) no internal relations. Existing assessments of atomism have focused almost exclusively upon the question of building blocks. On such assessments, the refusal of atomism amounts to entertaining instead the proposal that every entity has parts, and smaller entities are comprised of smaller parts still—a proposal that Jonathan Schaffer (2003, 512–13) refers to as *infinite descent*, according to which there are no indivisibles:

> What would a metaphysic of infinite descent look like? The most striking feature of an infinite descent is that *no level is special*. Infinite descent yields an egalitarian ontological attitude which is at home in the macro-physical world precisely because everything is macro. Mesons, molecules, minds, and mountains are in every sense ontologically equal. Because there can be no privileged locus for the causal powers, and because they must be somewhere, they are everywhere. So infinite descent yields an egalitarian metaphysic which dignifies and empowers the whole of nature.

According to this teaching, science—Physics, more specifically, or its successors or offshoots—will continue to discover and identify entities on increasingly smaller scales, as further decomposables. There will be no end to these discoveries.

However, a refusal to rule out internal relations (in other words, a refusal to accept the "no internal relations" plank of atomism) leads to a different conception of fundamentality, whether or not one adheres to the notion of building blocks, and so it is independent of the question of infinite descent. The rest of this chapter is devoted to fleshing out this second conception of fundamentality—*systemism*, as I shall be referring to it.

The antithesis of "no internal relations" involves adoption instead, as the fundamental unit of analysis, of the notion of a *System* rather than the notion of an independent entity. This is ultimately where countenancing internal relations will lead. What is a System? The notion of a System, with a capital *S*, is what the remainder of this chapter will be about. Adopting systemism will, appropriately, involve adopting a different vision of the relations among the sciences—and correspondingly a different theory of science. Importantly, this vision of the sciences does *not* simply solemnify the parody of nonreductivistic metaphysics, however attractive, that radiates from Augustus De Morgan's (1872, 377) lovely little verse:

> Great fleas have little fleas upon their backs to bite 'em,
> And little fleas have lesser fleas, and so *ad infinitum*.
> And the great fleas themselves, in turn, have greater fleas to go on,
> While these again have greater still, and greater still, and so on.

There is something profoundly lacking in this—decidedly antireductionist—vision of the order of the sciences. For it is simply false that there is biological order or organization on the nanoscale. And it is simply true that regimes of organization, order, and regularity, while not exactly following an order of scale or size, are very much constrained by scale. (Bonner 2011 arrays many reasons for why this is so.) To explain these facts should be a priority of any theory of the sciences. It is worthwhile, therefore, in service of a theory of science, to focus on the importance of scale, and the extent to which disciplines (biological, social, etc., and different, potentially competing theories within them) span the range of scales and are constrained by scale. These are neglected areas of inquiry.[6]

4. Systems

A System is not simply a network of independent entities standing in some random set of ("external") relations to one another—at a taxonomic level superordinate to that in which its components are rooted. A System is, rather, an assemblage or network of entities that stand in "nonelective" bonds to one another. These are the *internal relations*. What I mean by the term is this: it is not possible to assemble the entities in question, *as a System, without also* the bonds figuring in; it is not possible to assemble the entities in question, *as a System*, with arbitrarily chosen bonds. The bonds in question are ineliminable to the confederation of these entities, even if some of these entities can confederate with different entities, to form a still different sort of System. The relations between the entities in question are thus "internal" to their confederation. The bonds are therefore also an ineliminable aspect of an apt analysis of their Systemhood.[7]

This idea—that structuring relations are what make a System what it is, and so swapping out elements that can be swapped out without changing this structure would leave the System unchanged—this idea is basic to the theory of Systems. It is why Systems theory, which is a subdiscipline within the discipline of engineering, is a study of dynamics—of behavior—rather than of primarily static features. For on the Systems view, behavior is not "external"—not simply an accident of circumstance. Rather, behavior is characteristic; behavior is defining, because the relations among entities within a System are fundamental to its Systemhood. This is an alternative conception of fundamentality—an alternative to the building-block conception of fundamentality.

[6] Patrick McGivern (2008) makes a nice contribution to this small literature.

[7] Consequently, it will not always be possible to apply a standard causal path analysis to the interior workings of a System proper, since this analysis always assumes the probabilistic independence of contributions from different entities and factors—a Markov condition upon them. The description of internal relations in the text defies the Markov condition. For more on this see the next chapter.

We can illustrate this contrast in relation to the phenomenon of superconduction. Remember that superconduction is explained by Cooper pairing of electrons that—under the right circumstances—pair up; and each pair behaves like something neither member of the pair could be alone. So the Cooper pair has a property (the property of being a boson) that not only is radically different from the properties of its constituents (they are fermions), but also in some sense a contrary to the members' own characters—being a boson is contrary to being a fermion since the same entity cannot be both. But a pair of electrons, when properly bonded (when Cooper-paired) can defy any "logic" that might be local to the constituents. The bond (the Cooper pairing, in this case) is fundamental to the new logic governing its behavior in relation to entities at the new scale of interaction, and the logic that governs the behavior of fermions (the lower-scale entities) no longer applies. The bond is defining, and the identities of the constituents are irrelevant.

We need one important point of clarification before we proceed further. Not every aggregation of entities is a System with a capital *S*. A true System is defined by a clear *boundary* between "inside" and "outside". A boundary comes into existence when a critical mass of *control* is exercised, within a given region, on a number of control *variables*. I don't have space here to discuss the means by which this control is achieved (but see Thalos 2009). Suffice it to say that when control is achieved within a space-time region, certain macro states of an aggregation of entities that comprise a System S will form a set of interconnected equilibrium conditions that can be adjusted directly by means of making alterations to a (characteristically macro) feature of S.[8] Cybernetics, consequently, is the study of the properties of networks of interconnected equilibrium conditions.[9] It comprises precise treatments of the performance of control apparatuses and features special attention to defective behavior that tends to bring about oscillations (from mild to violent) in control-relevant quantities, when a control apparatus is poorly designed, mishandled, diseased, infected by a foreign control apparatus, or simply overloaded.

Adoption of a System conception, then, lays the groundwork for a proposition to the effect that situated entities (entities in situ), in addition to potentially (though not invariably) retaining a certain amount of independence, may also coalesce, in the process forming *molecular* Systems that further interact in complex ways with each other, as well as with and upon the atoms that compose them.

This conception of a System, according to which molecularization is the fundamental process, renders the bond itself fundamental to an entity's reality. Indeed it

[8] On my conception of control, no controlling entity is required in order for control to exist. A System under control is not necessarily a System under control *by* someone or something.

[9] Norbert Wiener originated the term "cybernetics"(from the Greek term for steersmanship) as a label for Systems analysis and contrasted it with Newtonian causal analysis in his (1948) groundbreaking opus *Cybernetics*.

renders the bond itself an object to which fundamentality may attach. From a Systems perspective, bonding is *the* fundamental entity-building process. In fact, from a Systems perspective, bonding is the *most fundamental* character or feature of (nonatomic) objecthood. Objects spring into existence as a result of many such (often overlapping) bonds: from the Systems perspective, an entity *just is* a thing bonded. It is possible, on the Systems conception, that independent atoms exist, entities that possess no internal relations. If this is so, it would be a completely contingent fact; whereas, by contrast, atomism simply assumes that it is a necessity.

Thus the contrast between atomism and *systemism,* as I shall call it, is the contrast, in ontology, between atoms in external relations (on the one hand) and situated entities (on the other). Situated entities are the subject of *dynamical Systems theory,* that brainchild of engineering science. From a Systems perspective, it is just as likely that there are internal relations "all the way down" as that there is a "ground" over which everything is constructed. Systemism is compatible both with atomism and with infinite descent. It is an alternative construal of what fundamentality itself consists in.

Another way to put this basic axiom of Systems theory is to say that the defining trait of a System is that it is a *unit of aggregation.* In language we introduced at the very beginning of the book, a System results from the application of a *sheath.* To appreciate this idea, it helps to bring to mind the engineering milieu from which it derives. Dynamical Systems theory is nowadays ubiquitous. From engineering (its point of origin and natural home), to physiology, and from psychology to ecology, it enjoys surprisingly wide application. Sometimes the analysis rings decisively false—as, for example, when adopted in certain treatments of historical narrative;[10] other times it is provocative and controversial, as when applied to the phenomena of mind and cognition.[11] Dynamical Systems analysis (or "Systems theory" with a capital *S*) is a strategy of analysis whose defining feature is its rejection of atomism. It mobilizes the language and mathematical technology of differential equations and brings into play the distinctive concepts of *equilibrium* and *attractor,* as well as *gain, coupling,* and *neighborhood,* that are not obviously proprietary to any particular domain of objects or regime in the world.[12] It is the ecumenical language of engineers, universal in scope. The unrestrictedness of the scope of Systems theory—the fact that it ranges over

[10] See for example the reaction of Roth and Rychman (1995).

[11] See books with such titles as *Stairway to the Mind* (Scott 1995), *Thinking in Complexity: The Complex Dynamics of Matter, Mind and Mankind* (Mainzer 1994), and *Creativity and Cosmic Consciousness* (Sheldrake et al. 2001). There are also less ambitious treatments, but nonetheless in the same spirit: e.g. Bedau (1997).

[12] Good introductions to the concepts and language of Systems analysis outside philosophical contexts are Abraham and Shaw (1992) and Broer and Takens (1992). And the entire December 2007 (vol. 37, no. 4) issue of the journal *Philosophy of the Social Sciences* is dedicated to treating it.

every scale of measurement—is key to understanding its founding ideas. It is as aptly applied to the social sphere as to the nanosphere. And this renders its concepts cardinally important to metaphysics.

The defining characteristic of a System in an engineering context is simply that it is a real-time construction built by a process—and so is subject to being added onto in the same manner. A System is essentially an integrated thing—a thing whose elements are integrated through organizational structures or protocols. Thus, whereas atomism adopts the point of view that wholes are *aggregates*, and thus not basic or fundamental, systemism adopts the contrasting view that *some* wholes are *integrates*, aggregates *clothed* or *sheathed* as a unity, and that for these wholes their confederation is basic to them and thus fundamental. This is the core contrast to which the rest of this chapter is dedicated to illuminating.

Systemhood is a phenomenon of integration; it is thus a phenomenon of *scale*. By this I mean that Systemhood is that phenomenon upon which the very fact of multiple scales is possible. Systemization, in the sense just defined, is the realization of capacities for *bonding*, and thus for further *up-scaling*. Bonding is a System's elemental or fundamental characteristic. The fundament of a System is not a ground or building block; it is a bond that raises the bonded into a higher scale of existence. A System is a One that is made out of many.

Simon (1973) made remarks suggestive of this conception of Systemhood. He thought that certain features of certain Systems could not be captured correctly by analysis of their parts taken together with the specific and heterogeneously conceived interactions among them. His idea derives from his concern with building complex things. The idea is that to build a complex System (as nature does), one proceeds in stages, with the result that at the end of each stage, what is constructed must possess a stable enough structure so as to "hold still" while the next phase of operations is being launched. Without these intervals or layers of stability, complexity (according to Simon) is unsustainable. This makes complex Systems typically (*a*) modular, (*b*) intersubstitutive in their parts, (*c*) qualitatively similar with a change to their parts or their number, and (*d*) stable under reaggregations of parts.[13] Simon thus grasped for the idea that a System as such has to be governed by high-order structures of stability (something that might reasonably be referred to as its *internal* relations) that are relatively independent of the sorts of interactions (physiological, chemical, mechanical, or what

[13] Wimsatt (1976, 2006, 2007) makes reference to many of these conditions in describing the difference between aggregates (on the one hand) and composed or evolved Systems (on the other); and Bechtel and Richardson (1992) illustrate the ways in which scientific methodologies attain what they refer to as emergent phenomena utilizing these criteria. This concern for layers of stability in complex Systems comports very well with the notion of control we have been at pains to articulate here. Indeed, I would insist that Simon had just this notion of control in mind when he talked about construction stages.

have you) that govern proceedings among their parts more locally. Relations of stability can be realized by any of such proceedings, or combinations of them; but their true significance is *as* relations of stability. A true System (with a capital *S*) is something special indeed and subject to high-order relations among its constituents. A system that does not obey such laws is one that quickly falls apart; it is no true System. This amounts to saying that a System, with a capital *S*, is one in which the aggregation of the parts has undergone a *reduction in degrees of freedom*. An illustration of how this happens is worth extended scrutiny, and we will turn to an extended example in this chapter—the case of flocking that we touched on several times already. Chapter 7 will be devoted to articulating and interrelating the technical concepts we have been deploying in the last several chapters (most critically, the ideas of *degrees of freedom* and *independence*) quite explicitly.

Many biological Systems, on an ecological scale, fall unambiguously (and unapologetically) within the scope of a Systems analysis. Consider the flocking of birds. This phenomenon is worthy of science. Only recently have dynamicists come to appreciate—and this largely through mathematical and computer modeling—that the remarkable synchrony of motion among flocking organisms can be orchestrated by simple adjustments of an individual trajectory, made locally but distributed widely. These can be understood as "all-over" features of a flock. Dynamicists now refer to these as "rules of flocking." The assumption is that these can be implemented via use of sensory modalities such as vision, sound, pressure, or odor detection. Assuming that each organism in a flock can sense local flockmates as well as its environment, and adjusts its own motion on an ongoing basis, Craig Reynolds (1987) for instance devised a computational model of flocking "boids" based upon the following "rules of engagement":

1. Separation: steer to avoid crowding local flockmates in your "near neighborhood."
2. Alignment: steer toward the average heading of local flockmates.
3. Cohesion: steer to move toward the average position of local flockmates.

If the theory of local adjustments is roughly correct, flocking requires uniformly of each flock member only that it react to flockmates within a certain small neighborhood around itself, characterized by a *distance* and an *angle*, as measured from the organism's own vector of motion. A flockmate outside this local neighborhood can be ignored—which is important from a bird's resource perspective.

Specifics of the large-scale features of the synchronized motion matter a great deal—because flocking needs to be very fine-tuned if it is to serve the interests of the flock. So how, precisely, does fine tuning of the "rules of engagement" scale to collective properties of the larger motion? For example, how does adjustment of the size of the neighborhood affect sensitivity to environmental features? How does it serve in location of food and avoidance of predators? The answer might surprise: much depends upon the specifics of the flock's "rules," and nothing at all depends on any feature of any given organism in the flock. The "rules" concern "all-over" characteristics that might well be shared with flocks of very different species, and may in no way depend upon the biology of the species. They are purely relations between the organisms; they are in no way tied to the characteristics of the flock members as carbon-based, oxygen-metabolizing entities. Furthermore, the rules are the flock's way of reducing its constitutents' degrees of freedom when in flight, and thereby modulating flock characteristics in a coordinated way. They are, in other words, constitutive of flock characteristics.

We can illustrate this reduction in the context of our present flocking example very simply. Close behavioral *coupling within a neighborhood of sensitivity* allows a localized change in direction to be amplified and propagated across the flock. This allows each flock member to influence and be influenced by flockmates much farther away than its local neighborhood—it gives each a much larger reach, a much more appreciable "effective perceptual range" than any one organism's actual sensory range. Study of the details of the scaling relations reveals that it is hard for groups to maintain cohesion if the *coupling distance* is too short. Longer-range transfer of information is enabled by increasing the coupling distance. Increasing the coupling distance further still creates a cohesive group, but "misinformation" can be more easily propagated, so the flock is more susceptible to error (false positives as well as false negatives). In contexts in which high sensitivity to motions of distant flockmates is unnecessary, and with it the short-term response to abrupt change in an environment variable, its implementation becomes inadvisable, given that it exposes the flock to unnecessary or undesirable fluctuations.

In the event that coupling can be adjusted to context, context-flexibility can be a highly desirable flock characteristic, but it has costs. For example, if individual boids in flight can condition their reactions upon context (under threat, for example, aligning more strongly with distant flockmates, increasing "System gain") when the environment is volatile or dangerous, this could allow for some flexibility in response time. But increased System gain creates at the same time a heightened sensitivity to weak or ambiguous environmental signals, thereby increasing susceptibility to false positives, where by contrast damping response to local fluctuations in less risky environments increases false negatives. A balance has to be struck for each ecological condition. It is an evolutionary challenge.

What becomes more and more clear as dynamical analysis of flocking proceeds is that a flock of hundreds of organisms, operating under a set of "rules of engagement," is decidedly *not* a System with degrees of freedom on the order of the number of flock members: it is instead a System with something on the order of a dozen degrees of freedom (*levers*, if you will), counting among them rough size, coupling distance (size of the neighborhood of sensitivity), and level of context-sensitivity, as well as environmental variables that tend to couple with these features. From a flocking perspective, a flock is an entity with a reduced number of degrees of freedom. And some of these degrees lie in the environment itself! These degrees of freedom displace many or most of the "micro" variables that predominate—that advocates of atomism would purport as relatively more fundamental—when the boids are not flocking, as soon as the flock members begin governing themselves by the rules of engagement.

In this analysis there is a transition from individual organisms operating with "rules" to flock characteristics that determine the way the flock navigates the environment. The "flock" comes into existence as an entity, as a "natural kind." Is this analysis biological? Does it belong in biological science? Or does it belong in some "higher" level? Does flocking belong to some proprietary level of reality instead? Does the answer to the preceding questions have any real significance? Finally, does the flock, as such—as something over and above the members—really exist?

Examples like the boids demonstrate in fact that so-called levels enjoy no true existence in the world. They simply demarcate lines of temporary convenience. But the discussion of the different conceptions of fundamentality illuminates cardinal questions in metaphysics about natural kinds and their relations to constituents—in particular, it illuminates the fact that properties of parts do not settle properties of wholes in any straightforward way, so that a theory of natural kinds (if there are natural kinds at all) is no trivial business.

This fact prompts some important questions about natural kinds, in connection with the contrast between synced and linked models. These models are not contraries, since a linked model can also be a synced model (as we illustrated in chapter 1). Still, one can ask: is it better to articulate the theory of natural kinds utilizing the concepts of linked models, or synced ones instead? We will return to the topic of natural kinds later in this chapter.

5. Displacement, not Reduction

From a Systems perspective, what transpires from one scale to the next, when new features appear and others move aside, is just that—*displacement*. Reduction, in the contemporary physicist's sense (which, recall, harks back to Carnap's conceptualistic ambitions), is inapt. Nothing is eliminated. Coupling distance simply appears as relevant, and subsequently—so long as the System persists as a System—dominates over features of the flock that would be more appropriate outside the flocking context. But coupling distance is never "reduced" to anything more "basic." Purportedly "more basic" features (i.e., birds and their individual locations, as such) are simply sidelined. They do not disappear; but neither are they relevant for the purposes of the flock and its life. When birds flock—when we have a System of boids—coupling distance (neighborhood of sensitivity) is a factor not to be ignored when explaining what transpires. Purportedly "more basic" features of the birds as such are irrelevant to flocking rules. They are simply ignored. Their relevance is recovered when the birds are literally on the ground; there, coupling distance is not needed to explain their behavior. The boids are simply birds then. And that is that. That is how Systems analysis proceeds.

This procedure reveals something important about Systems *as such*. Systems come into existence with bonds, and they de-Systematize when these bonds dissolve—as many Systems (like flocks) are very much apt to do. Reduction in degrees of freedom is key to the process—it is indeed *the key metaphysic* by which Systems are built. ("Metaphysic" is the better term, for the word "process" does not rise to the correct level of abstraction.) These reductions are by no means the reductions spoken of by atomists. These reductions are true reductions in complexity, a topic to which we shall return by way of concluding the chapter.

Thus what is fundamental to Systemhood is the bond. In strong numbers, bonds in situ can effect reductions in degrees of freedom. A sufficient number of reductions in degrees of freedom is key to the (manifest) *granulation* of things into natural kinds—the crystallization of order at a certain scale of measurement marked by the appearance of objects ("granules") of that length scale. An aggregation of entities is not a System if no such reductions take place. To be eligible for Systems analysis proper, a configuration of entities must be more than a mere aggregation.

6. Systemism Elsewhere

Use of the term *systemism* did not originate with me. Mario Bunge, a prolific philosopher with roots in physical sciences, is its originator. He first utilized the term (in the 1960s) in the process of articulation of a middle ground between methodological individualism, on one side of a certain debate in the philosophy of social science, and holism, on the other. The concept was revived in the 1990s in the context of a renewed discussion of so-called "mechanism," which has been taken up widely since. The "new mechanism" appears in the philosophy of biology (with Bechtel and Richardson 1993, Craver 2001, Machamer, et al. 2000) as well as once more in the philosophy of social science (Elster 1989, Hedström and Swedberg 1998, and Little 1998).

Bunge receives some credit for the notion of systemism,[14] but it is clear that those giving credit do not believe that Bunge's articulation makes it clear how systemism is any kind of clear middle ground between two poles (individualism and holism). Keith Sawyer (2001, 2002), while tipping the hat to Bunge, puts forward his own articulation of this same would-be middle ground in terms of the notions of supervenience, multiple realization, and emergence. Now my articulation of the term "systemism" is different from Bunge's—and Sawyer's too—although I do believe it is what Bunge intended. Let me explain.

Early Bunge (1979) maintained that a System consists of an ordered set (therefore, an abstract object) of (1) the component entities and (2) the set of relations in which they stand. Crucially, this definition of a System does *not* define an *object* with properties in a class distinct from those that characterize component objects. This characterization also skirts entirely the issue of whether the relations are internal or external, and so does not (as Richard Langlois was quite right to point out) provide an intermediate point between holism and individualism. True: a serious holist would find it objectionable. However, "What is *not* true is that systemism somehow represents a new methodological alternative: the basic ideas of what Bunge calls systemism are essentially identical to what sophisticated methodological individualists have believed all along" (Langlois 1983, 586).

In the 1990s, Bunge began to speak the language of mechanism and process. And most recently (2004, 188), he has arrived at the following: "a system is a complex object whose parts or components are held together by bonds of some kind." He is thus now tending in the direction of "internal relations," though he has some way to go.

Now my approach, like Bunge's most recent efforts, is fundamentally ontological, nonabstract (speaks of objects rather than sets), and does not speak of theories, the laws at which they gesture, or how we come to learn them. And, unlike Sawyer's, my approach does not rely on a ready-made (but seriously flawed) metaphysical doctrine, such as supervenience, against which I have already lodged my criticisms in the early chapters of this book. Rather, and again like Bunge's, my conception seeks to articulate principles that ground dynamical Systems analysis and the ways engineers especially apply it to real-world Systems.

Finally, my articulation of systemism *really is* a middle ground between atomism and holism. For it has these three properties: (1) it rejects atomism; (2) it does not ignore the (material) entities and relations that hold them together in

[14] Two special issues of the journal *Philosophy of the Social Sciences* (2004, nos. 2 and 3) were dedicated to Bunge's contributions to the social sciences by way of his defense of that general idea.

a System, allowing them to interact with one another and with the System's more macro features; and (3) it speaks of the bonds that hold parts together as themselves fundamental, indeed as internal. And, despite (1)–(3), it falls shy of two objectionable (or at any rate objected-to) features of holism: (1) it does not define the parts in terms of the whole or its "functions," so that parts do have an ontological reality independent of the whole; and (2) it clearly acknowledges the difference between simple aggregates and true integrates—only the latter of which are true Systems.

So whereas holism might "seek to describe, to understand and in most cases to explain the orderliness and stability of entire social systems . . . [and insofar as it treats] individuals, the treatment comes after and emerges from analysis of the system as a whole" (Barnes 1995, 37), complete with characteristic "functions" such as the production of goods and services, systemism makes no such sweeping assumptions. Instead, systemism requires justification of each and every application of a Systems analysis to a set of social institutions (requiring demonstration of the propriety of employment of its defining concepts—the notion of boundary, for instance). For it also recognizes that some aggregates are not integrates in the relevant sense. And it recognizes too that even integrates do not enjoy "functions" in any defining or proprietary manner.

7. Different Theoretical Linkages

Systemism as articulated here—namely, as a construal of fundamentality that contrasts and competes with the building-block construal—has resonations rather different from those of atomism. It is linked with an entirely different range of theoretical issues. For systemism is intimately and uniquely positioned to draw, within the parameters of scientific metaphysics, the very notions of *definition* or *essence*, and so of discipline-specific natural kind terms. These are rarely called upon these days to play key roles in *organization* of the special sciences, however eligible they present themselves for the job. (As I've already remarked, "levels" are specified with reference to disciplines—for example, the "physical level" or the "biological level." Levels are thus dependent, at least for definition, upon the administrative units typically found in universities—hardly inspiring as "fundamental" categories.) But when natural kinds are in some way fundamental, they can play a role in the founding of disciplines. Key questions in the philosophy of natural kinds, if drawn into the orbit of fundamentality, will become more lively. Here are a few examples.

Once the questions concerning internal relations are posed as key to explorations of (at least some) questions of fundamentality, and it comes clear that the defining question on the topic is *not* "Is there a fundamental level?" but rather "Are there internal relations?" the link to certain core questions in the philosophy of science will be open to engagement in a new way. Because atomism is now the reigning view of fundamentality, the reigning view on the subject of nature is that there are no fundamental natural kinds outside of Physics. Consequently questions arise as to the true autonomy of nonphysical sciences. Now, I have sought here only to establish a competing conception of fundamentality, and not to argue for the fundamentality of any specific kind, entity, or feature as fundamental under the new conception. This is the space that must now be mapped if we are to establish the autonomy of any nonphysical science, including, of course, biology, psychology, and sociology. But it becomes clear that there are important links to be forged—or refused—between conceptions of fundamentality and conceptions that treat other questions (for instance, the autonomy of different sciences).

A critic of systemism as a construal of fundamentality might inquire as follows: why suppose that the core question to which systemism is one answer (the question whether there are internal relations) is really a question about fundamentality at all? Why isn't it an entirely orthogonal question? Of course this query takes us to the very heart of our concern for fundamentality. What work does the notion of "fundamental" do in a conceptual economy?

But the proponent of systemism as a legitimate construal of fundamentality will have available a straightforward reply to this question: systemism is an eligible answer to the question of what is fundamentality because systemism proclaims that there are order-establishing entities that are not themselves atoms, but rather bonds or unifiers—adhesive, not substance. The question of internal relations pertains to the foundations of order. And this makes the question to which systemism is an answer worthy of the name of "fundamentality" simply on the basis that the concept of "founding" concerns orders of existents as to kind or category. (Remember Cooper pairing: the bond between electrons in a Cooper pair is a determinant of the category of object—whether fermion or boson—of the resulting entity.) Systemism addresses the question of the possibility and origins of a One out of the Many. Surely this concern makes it worthy of the name of a theory of fundamentality.

Answering the question of whether there are natural kinds outside of Physics is not an immediate objective of this book. The immediate objectives are more foundational and include illuminating the fact that questions regarding natural kinds resonate intimately with certain questions concerning fundamentality construed in the way here articulated, because these questions are intimately linked to issues concerning the difference between internal and external relations. There are many issues to explore while addressing these matters that we cannot enter into in any great detail just now—such as (in no particular order):

1. When a new sphere of order is created, are there other things, in addition to internal relations, that do the creating? Laws of nature, perhaps? If laws are required too, are they more fundamental still than the internal relations?
2. Do internal relations create other things in the new sphere of order?
3. How are internal relations (if there are any) related to the "laws" of the new order (if these exist)?
4. Is there more than one set of possible internal relations for any fixed set of aggregants? What happens in such cases? Do the aggregants exercise any freedom in choosing the regime that will govern them? Does the outcome depend on individual features of the aggregants?

8. Complexity

We have been told that we need Systems analysis because the world is complicated. Because there are far more "basic" entities, more bodies, more particles, and hence more quantities than a mathematical model founded on "fundamental" physical laws can handle. Moreover, things are supposed to get many orders of magnitude more messy as we broaden the scope of concern to topics outside of physics. Some theorists celebrate complexity—and see it as the basis of the "emergent"—if there is any such thing. We saw in earlier chapters how certain philosophers invoke complexity in support of emergentism, for example S. Mitchell (2009, 44): "Taking account of nonlinear dynamics and feedback causal process in addition to static and linear representations that may be adequate for simpler domains of nature is one of the ingredients of an integrative pluralistic ontology and epistemology." And Batterman (2002) presents a narrative of emergence as asymptotic reasoning, but also grounds the analysis in a notion of complexity.

Rather than celebrate complexity, another corps of thinkers sees it as a barrier or roadblock—it stands in the way of a rigorous accounting of things, necessitating a certain (and quite deplorable) laxity. They present the necessity for a less fastidious manner of analysis as a matter of practicality—an "applied" rather than a theoretical matter—not a matter of principle. Systems analysis is thus, according to this idea, a concession to human frailties (for instance: Teller 1992). We stoop to Systems analysis—where we do—because it is a simple and incontestable fact of life that only superhuman intellect or still-unattained mathematical facility can handle the computational complexities we face when we attempt to treat in theory the many-entity Systems we must navigate in real life. Complexity is, on this conception of things that is characteristically pragmatic, an occasion for being practical—for yielding to a lesser form of analysis.

If our discussion on the topic of fundamentality is right, however, this assessment must be rejected. For if our discussion is correct, then Systems analysis offers a *competing* analysis of complexity itself, as well as a competing analysis of fundamentality. If our discussion is correct, we must reply to Systems detractors as follows: we need Systems, *not* because the world is so complex (though it is frequently complex) but because it quite frequently *is not*. For Systems does not oversimplify what is *genuinely* complex. Rather, as we have seen here, Systems analysis is capable of taking advantage, in an elegant way, of certain serendipitous *reductions in that complexity*. For if systemism is true, then we must amend the atomist's conception of the world in the following way: in addition to possessing structure at smallest scales, the world also possesses fundamental structure also at higher scales. We must, in other words, accept that our world is possessed of layers of order. This is the order that Systems analysis is designed to capture. Systems analysis gives us a general, recursive analysis—that is to say, a meta-analysis—of how order at the next scale is structured, in neutral dynamical/metaphysical (not special-scientific) terms. Systems analysis is thus the metaphysics of higher order. And it rests on an alternative conception of *both* fundamentality *and* complexity.

Consequently, to the critic of systemism we are bound to reply: we require Systems analysis because there is more order—and correspondingly *less* complexity—in heaven and earth than is dreamt of in the philosophy of atomism. Such a metaphysical platform of fundamentality is a more firm foundation than we have had to date for a theory of science that aims to accommodate each special science as autonomous.

9. Will the True Conception of Fundamentality Please Stand Up? The Theory of Science, Part I

So there are at least two eligible conceptions of fundamentality. The function of the first (atomism) is to stop us asking why at a certain, absolute point in the process of inquiry. The second allows us to found new sciences, without suggesting that they must be dependent, or at least related to, a "more fundamental" science via a dominance relation of some sort. Which is the better conception? Is there such a thing as better? It would seem that I am partial to the second conception of fundamentality.

This is not the case. I view both as important, but also as orthogonal—they do different work, and we must not confuse the service rendered by the one for that rendered by the other. To illustrate: it is important to inquire (1) Are there independent atoms governed by "fundamental laws"? It is a question to which we might never have a conclusive answer. My money is on the negative, because all the physical entities we know of, with the possible exception of light (and it's not a building block of other entities), are entities-in-context (a proposition I put forward in chapter 2). But even if we cannot be confident of this answer to (1), it is still important also to inquire (2) What holds concrete entities together at whatever scale they happen to inhabit? The answer depends on the scale of the universe—on the answer to the question of whether there is activity at higher-than-smallest scales that contributes to their holding together. (1) and (2) are genuinely independent questions. (2) is the question to which systemism is the answer. Systemism is thus an answer to how objects manage to hold together under certain classes of circumstance, even if it were true that there are atoms in the universe.

Someone may now object that the point of looking for a notion of fundamentality has got to be different from the ones I have thus far articulated. The point of looking for a notion of fundamentality is that we're looking for the "cement of the universe" in the most general sense—the most basic thing that holds the universe together. Surely that's a worthy quest. And equally surely, there's a quarry, and so also a right and a wrong way to answer the question.

That someone may go further and insist that we already have hold of that conception: it is the conception of causality. Causal relations are what hold the universe together. And philosophers have been onto it since time immemorial, however badly they may have botched the job of articulating it and spelling it out. But anyway, that's the project.

In the next chapter I will be arguing that causality is not the cement of the universe. More specifically, the language of causation is not a lingua franca among the sciences. Furthermore, the language of causation is much better construed as the language of how certain factors, in a population of cases, raise the probability of certain other factors also being manifested in the population (for example, how certain genes in a particular population of organisms raise the probability of certain behaviors or physiological characteristics). If we grant the value of using the language of causality for this "probability-raising" conception, it will be ineligible for a more general usage as a lingua franca. And that's why we require a conception of fundamentality instead, for the sake of unifying the sciences.

Some readers will think it's just fine if there is nothing to unify the sciences. (And so whatever services the language of causality renders, even if it isn't a lingua franca, are sufficient.) Why should we expect the sciences to be unified anyway? If we have given up on the idea of a Master Science (as I've said we should), then what is to recommend unification wrought by other means? I said at the very outset of these investigations that the sciences do something that the arts do not: they speak truth in a way that everyone with an opinion on the topic in question has to listen to. Science speaks with an authority. That must be something that all the sciences share. And since that starting point in the introduction to the book, we have added that no science enjoys more authority than any other—that the sciences thus enjoy equal dignity in relation to one another. This is the idea that there can be no hierarchies among the sciences, since the universe is active at all measurement scales. If all of that is correct, it cries out for explanation: why should it be so? And if, moreover (as I shall argue in the next chapter), the language of causality does not enjoy the same status as that of science, then in virtue of what do the sciences deserve the dignity they plainly enjoy?

I answer that the same thing that explains the unity of science also explains the dignity that is its due. It is the function of science to throw light upon the phenomena of the world and everything in it, and in the process to make them intelligible and effable. Herein lies its dignity. In this regard it is distinct from causal analysis. I like to think that in taking this stance I am siding with David Hume—the famous problematizer of causation. As I read him, Hume is skeptical of the value of causal analysis because he is extremely sensitive to the fact that the language of causality (which, incidentally, we have come to utilize in just the way Hume did, although he did more than anyone else to prompt new conceptions of the idea) is profoundly uninformative, profoundly small. Hume was prepared to grant that Newton's laws of motion—framed in terms of forces and the motions they impart to bodies upon their communication—are an accurate characterization of causation. But he was relentless in asking by what *powers* (as he put it) a thing exerts a force. At one point in his magisterial *An Enquiry Concerning Human Understanding* (2009, 562 n. 18) he considers the law of inertia:

> I need not examine at length the *vis inertiae* which is so much talked of in the new philosophy [Newton's *Principia*], and which is ascribed to matter. We find by experience, that a body at rest or in motion continues for ever in its present state, till put from it by some new cause; and that a body impelled takes as much motion from the impelling body as it acquires itself. These are facts. When we call this a *vis inertiae,* we only mark these facts, without pretending to have any idea of the inert power.

It seems that Hume expected, or at any rate sought, an illumination that Newton's theory of dynamics does not furnish. I thus read Hume as saying that Newtonian dynamics, understood as an analysis of causality, is not worthy of the dignity of knowledge—of the dignity of Science. I like to think that he would have thought differently of quantum theory—or at any rate, of applications of quantum theory to the structure of matter, and in its gestures at the duality of wave and particle.

Construed purely as an analysis of the probabilistic notion of causation that we share with Hume (and more on that in the next chapter), Newton's theory is indeed a sorry excuse for science. The best way to defend this point is via example. Consider two contrary hypotheses on the subject of obesity. One hypothesis (touted as fact by a preponderance of the scientific community) is that obesity results from ingestion of an excess of saturated lipids, which in turn has the effect of causing numerous diseases like diabetes and heart disease. A contrary hypothesis (maintained by a slim minority) is that obesity and the associated diseases (and others besides) are common effects of a different cause, a deficiency in the metabolism of sugars and refined starches: this metabolic problem is the common cause of the other ills, including obesity. These are distinct hypotheses—since obesity in the latter hypothesis is an effect (on the metabolic deficiency hypothesis) and not a cause of the named diseases. If one is true, the other is false. And surely we care—and have need to know—which is the true hypothesis. It is absolutely incontestable that there is value in the knowledge. But notice this: neither hypothesis, even if true, sheds any real light on the relevant physiology. A biochemical theory that painstakingly details the operations of saturated fats and sugars, why they have the functions they do in the body normally, and therefore which hypothesis is correct, might shed such light. And such a theory *would* count as Science. I like to think that Hume would have joined me in saying so, and in recognizing that biochemistry is due a reverence that neither causal hypothesis, on its own, merits. He would not have thought biochemistry quite as scientifically inadequate as Newton's dynamics.

In point of fact, I think that Hume was wrong to believe that Newton's dynamics is small. This is perhaps because I see the rivalry between Newton's dynamics and the theory of analytical mechanics (which we discussed in chapter 2), on the topic of motion and its sources, as extremely illuminating vis-à-vis the nature of entities and their characters. But from where Hume stood, he could not appreciate this contrast.

By way of concluding this chapter, I will summarize the few shreds of a theory of science I have put together so far. (I will pick up these threads of the theory of science again in chapters 6 and 7.) Science illuminates the world and everything in it, at every scale of measurement. So a theory of science is a theory of knowledge—it is at least a theory of a very important expanse of knowledge. A conception of fundamentality in terms of how entities hold together and interact to form even larger entities will serve us very well indeed when it comes to articulating a very general theory of the world at its different scales (in a grand metaphysic). This conception of fundamentality will allow each science to speak its part of the story of how objects hold together, giving each its due and proportionate attention. This means that a metaphysic of the world has to pay attention to actual science. It cannot proceed exclusively from the armchair.

4

Why Causation Cannot Be the Cement of the Universe

The magi, as you know, were wise men—wonderfully wise men—who brought gifts to the Babe in the manger. They invented the art of giving Christmas presents. Being wise, their gifts were no doubt wise ones, possibly bearing the privilege of exchange in case of duplication.

—O. Henry, *"Gift of the Magi"* (1906)

1. Gifts of the Magi

Locality is venerated for a reason: it seems to encapsulate the notion that entities respond to causal laws—that things are governed by imperatives, and that the information required for the (proportionate) response must be "present at the same place and time" (the fine, zero-dimensional spacetime point) at which the prescribed response is called for. According to Newton's laws, for instance, one responds to the imperative of a force (a kick or a tug) by moving in proportion to the size of it and in the appropriate direction (say, according to Newton's Second Law of Motion). Locality is thus associated with a picture of law-abiding citizenship in the regime of the physical. It is the very emblem of causation. Nonlocality, by contrast, has no emblem. Because it is the antithesis of the idea of proportionate response to local action, common sense regards it as magic and miracle. Magic is the folk name for nonlocality. But note: nonlocality is present in (for instance) the idea that everything everywhere is responding all at once to the imperative to conserve the total sum of the energy of the universe.

The rationale for imposing locality as a requirement is intelligible from a certain point of view. Like imposing the condition of causal explanation itself, it seems to go hand in hand with banishing magic and mystery. But in the guise or office of imposing a requirement of nonmagical narratives, it leaves a great deal to be desired. To explain this point, I will discuss an example of a simple glancing elastic collision between two billiard balls. Consider a billiard ball (A) moving along a billiards table toward a second one (B) of equal mass and proportions, at a glancing angle. (See figure 4.1, the "before" panel). Because of this orientation, the billiard balls will scatter after collision—the first will not transfer all its energy and momentum to the second. This situation is one in which total energy is conserved. This is achieved through conservation of linear momentum—which requires only local information. When the collision occurs, the balls are to divide up the energy into two orthogonal dimensions at the point of collision—where all the information they need is situated. They are to ensure that the (vector) sum of linear momenta in each of these directions is preserved. Not a simple thing, for billiard balls. The information is all local to them, I grant you. But how to use that information to achieve the "after" configuration in figure 4.1 has got to be a daunting task if you're a billiard ball. So locality, all on its own, is not automatically the clear picture of nonmysterious response to imperatives that its friends suggest it might be. Why insist upon it in our paradigmatic examples of causation? Why does it make the narrative of our billiard balls more intelligible? Why should we regard it as more mysterious to demand the universe conserve energy than to ask two billiard balls to transact successfully the balancing act required to transition between "before" and "after" in figure 4.1? The answers to these questions are nowise clear. For to banish the nonlocal, as illegitimate by way of explanation, does very little or nothing to oust mystery.

The story is now routinely told that the balancing act is performed via a process of the kinetic energy (the energy of A's initial motion) being transformed into potential energy—an energy of compression, like in an ideally springy spring, along the axis between the balls' centers—and then transformed back into (proportionate) kinetic energy in each ball. This is a *model* of local action, manufactured as a companion piece to the (already local) description of the linear momentum conservation process. So a model in compliance with the imperative of locality was produced to guide the imagination in conceptualizing the process. Indeed the fact that we feel the pressure to manufacture the "spring" model is reason enough to suggest that locality, as a stand-alone condition, does little to dislodge the sense of mystery people experience on being presented

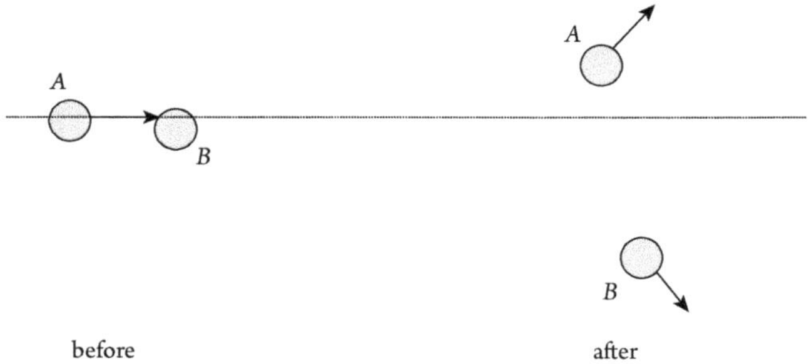

Figure 4.1 An elastic collision between two billiards

even with a local process. Mystery can remain alive and well and kicking vigorously—if no models are offered alongside to assist the imagination. And it is not in the least bit obvious that no suitable model will be available to dislodge the discomfort and mystery associated with *non*local processes.

In truth, the models that respect locality (on one hand) and the condition of locality itself (on the other) are distinct things. Let's suppose that we can always generate intelligible models for local processes. What's to prevent us *also* manufacturing intelligible models for nonlocal processes too, much as we produced the watermelon physics of the previous chapter in order to make lasers intelligible? So, why venerate locality?

Locality is the poster child of causation because its patrons require models in compliance with certain prescriptions. They don't accept models that fail *as explanations*. And so long as stories can be told in keeping with these prescriptions, locality is safe and the preferred conception of explanation is preserved. When this becomes impossible, we are (once again) without a picture of the preferred type. For better or worse, models of quintessential quantum processes can't be modeled in the preferred way (remember Bell's Theorem of chapter 1). But so what? The solution, as I said at the conclusion of chapter 1, is simply to innovate more models—like those of watermelon physics. We can't afford not to if a nonlocal theory is a perfectly good predictor of what we need predicted—and if (more importantly) it's the only game in town.

If we decide that it is useful to maintain such models as quantum theory might be able to help us devise for the purpose of generating explanations, then what do we do with our prescriptions of locality? These prescriptions, or so some thinkers have insisted, seem to go hand in hand with the notion of causation. One way of proceeding is to unhinge the two conceptions and allow causation to track the modeling we find in the sciences, whether or not locality too is tracking along. Causation can then be the lingua franca of the sciences.[1] This move has its costs: we lose the connection between causation and locality, and—more importantly—we will have no way of demarcating between causation construed as a local process, and causation construed otherwise. This is because causation would in that scenario name anything that can be articulated in a model or narrative.

I propose something different. Causation, as constant companion to locality, is already very well entrenched, thanks to its uptake in the social and medical sciences. (And as we'll see in this chapter, this conception is also very much in line with the probabilistic conception that has been well studied by philosophers.) So I propose we keep the name of causation associated with this conception (leaving it as an exercise for the keen philosopher to work out how it is connected with locality). As for what happens in a scientific model, we must choose another conception.

[1] This is Nancy Cartwright's move (1989).

Separating causality from what is done in scientific modeling is crucial if we are to utilize the notion of causation in a genuinely useful way. That genuinely useful way must part company with the language of science, as this chapter will detail.

O. Henry wrote a brilliant story about "two foolish children in a flat who most unwisely sacrificed for each other the greatest treasures of their house" (Henry 1982). In the story, each of these protagonists pawns a priceless personal possession in order to purchase an accessory for the other's possession. The result, of course, is that what each has at the end of the day is essentially worthless; they are collectively impoverished by their mutual generosity. We read this story on gift-giving occasions, perhaps in service of memorializing a quality of love, or just sentimentalizing gift-giving opportunities. But the story is more than that. It is in an oblique way really about the original Magi and how these kings journeyed in search of they knew not what, bearing gifts fitting only for a king but useless to someone who does not fit their conception of royalty. In some sense these gifts (as do those of O. Henry's protagonists) served a purpose, but not the one they intended. The Magi are allegorical of the human intellectual journey with the concept of causation—it is a cautionary tale vis-à-vis science and its relationship with causation. Too much generosity—in this case, generosity with use of the term "causation"—can leave us bereft in real terms. I will suggest that we should not be so generous with our conception of causation. For if, unlike the "two foolish children," our efforts are not to be for naught, we must be prepared to make exchanges where prudent. To save Science itself, the scientist must surrender the notion of causation to the clinic.

2. The Philosophical Portmanteau

What holds the universe together? Aristotle propounded a fourfold theory of why things exist, in which he described four types of causes. By the early modern period, only one of Aristotle's causes had survived with its original name of causation intact—the one that has since come to be known as "efficient causation," and in today's currency "actual" or "token" causation. This conception plays a role in Descartes' mechanics. And it is the guiding notion of causation at work in Newton's mechanics, which owes so much to Galileo, and which subsequently receives its highest expression in Einstein's Theory of Relativity. The pruning was a wise move, enabling the conception of causation to flourish after a sort of speciation event.

The guiding beacon in the modern conception of efficient causation is the notion of locality or local action. But in the beginning of the twentieth century Bertrand Russell predicted that causal concepts would be gradually supplanted by conceptions of natural law and lawlike correlations and would thus become superfluous, ineligible to play a role in the metaphysics of science (Russell 1913). Still, the conception of causation did not pass away. Hans Reichenbach argued strenuously for it as the core of the Theory of Relativity in the 1930s and 1940s. And in 1980 J. L. Mackie referred to causation as the cement of the universe—memorializing in "cement" its place in science. Recent decades have seen no waning of interest, but indeed a revival of philosophical concern with questions concerning the role of causal concepts in such diverse scientific areas of heavy intellectual traffic as biology, neuroscience, ecology, and the social sciences.

In the tradition of mechanics, the notion of cause is crystallized as "force." (Indeed Hume was quick to associate the term "cause" with the forces Newton began to speak of in their era. See Hume 2009.) There is room in modern physical mechanics quite generally for multiple sources and types of force—such diverse forces as, for instance, gravity and the electromagnetic force all have a place in physical mechanics. But there are physicists today who seek to reduce the number of forces to fewer and fewer, so as to be able to construe them all as manifestations of the same thing, or at any rate collaborators on a "unified field" whose result would be one and only one theory of physics. This idea has consumed the energies of many theoreticians for many decades now, not least of them Albert Einstein, without clear successes.

My potted history of the last several paragraphs speaks of an aspiration in natural and empirical science to simplify the world, in spite of—or perhaps it is because of—the complexities we discover in it. This is a metaphysical enterprise. Our contemporaries, many in the discipline of Physics, find simplicity compelling, a real guide to truth, whereas Aristotle's research program obviously suffered no compulsions to describe the world as simple. In this province too, we see displayed among contemporary physicists another instance of foundationalist cognition, of which we saw a related instance in the last chapter when we discussed the related topics of the building blocks of the universe. But the world and its causes might not be so simple as we might wish. There are no more assurances that there are fundaments vis-à-vis "causation" than there are vis-à-vis the furniture of the universe. There is no guarantee the world is so simple. Our sciences—if they insist on simplicity—would be in deep trouble if the world does not heed a simplicity imperative. And likewise Philosophy would be in trouble if it were premised on an unshiftable assumption of simplicity.

Still, there is a school of thinking about causation that is unconcerned with fundaments in this sense and seeks to identify the contours of causation in its barest conceptual terms, without regard for how it might actually be realized in the world—whether through simple or complex, single or multiple ways. The idea is to characterize causation abstractly enough, simply as a way of characterizing in an abstract way the nature of universal cement, making it possible to slot in later such details as physicists or others might turn up. It would be something like a portmanteau concept, a way of being able to say when something qualifies as a cause and so eligible of treatment in a particular way.

This is the project to be discussed in the present chapter. The conclusion here will be that there can be no such portmanteau conception that embraces all the projects of all the natural sciences—never mind social sciences or mathematics. The conception of causation that prevails today—and which is the best conception of causation, and one I wholeheartedly acknowledge as useful—is unworkable as a portmanteau conception. In this chapter we encounter the notion of *protocol*, which will help us formulate in a more embracing way the notion of *order*. Both will be useful in the subsequent chapters for elucidating further the preferred conception for a theory of science: that of dependence. These are the exchanges I propose we make, since as O. Henry's story teaches, it is wise to be prepared to make exchanges.

3. Stumbling Blocks on the Royal Road to Philosophical Causation

Imagine that two assassins are conspiring to take down a certain political figure. The plan (or, as we will eventually say, the *protocol*) is for assassin 1 to fire upon the victim (call this would-be event C1), while assassin 2 will act as a backup and fire (C2) in the first assassin's place should assassin 1 fail to do so. By bringing about C1 (and thereby the victim's demise E), assassin 1 *also* contributes to *prevention* of E, through acting as a cause of the nonoccurrence of C2. This apparent fact about causes—that a single event can function both as a cause of a second event and also as a preventative or protective factor against that second event—is a source of much mischief for numerous philosophical theories of causation. It seems no philosophical theory seeking to conceptualize causation in terms of simpler concepts, such as those of probability, is immune to the problem.[2] So the problem has received a lot of attention under the label of *preemption*.

[2] Preemption is discussed widely. For a recent discussion of the problems it poses for a counterfactual account of causation see Schaffer (2000). I am taking my example from work by Christopher Hitchcock. In a series of recent articles, Hitchcock has made much of a distinction between (on the one side) the *net effect* of a certain factor or event C on a second factor or event E, and (on the other side) the *component effect* of C on E *along a particular causal route*. (See Hitchcock 2001a and 2001b for supporting arguments.) Hitchcock argues that when C affects E in more than one way, this distinction is required to capture all our causal judgments about a given scenario. Hitchcock seeks to use this distinction to resolve the long-standing difficulties in the analysis of preemption-type cases. By distinguishing between the two kinds of effect had by C1, Hitchcock proposes to offer a more satisfying account of all the relations of causation involved in cases of preemption. (This problem structure, and closely related others, threaten David Lewis's counterfactual account of causation; see Lewis 1973 and 1979.) But Hitchcock's analysis, illuminating as it otherwise is, still leaves out a good deal that is relevant to how this E of ours (the assassination) is being brought about, what pertains to the control structure, as discussed below.

In the general structure of these cases of preemption, one cause (C1, for instance) serves two functions: (1) it directly brings about an effect E; but (2) it also prevents a second potential cause of E (C2) that was poised to bring about E had the first (C1) failed to do so.

Preemption, illustrated in our assassination example, is an instance of a genus of a larger problem faced by causal analysis within engineering contexts—the problem of modulation. Consider the hypothesis that water promotes plant growth. To be sure, sometimes watering a plant too much will promote its decay rather than its healthy growth; and too little is no better, but neither fact should be sufficient to displace the generalization as a true causal one. So causal laws are often construed as couched in (sometimes implicit) provisos that allow for a mean or average effect (and so we get the "exception that proves the rule"). The provisos spell out the details of a "background" against which the cause, when it appears, explains production of the effect. Suppose now (to stipulate the facts of a hypothetical situation) nitrogen levels boost plant growth only if accompanied by sufficient levels of moisture, whereas moisture always boosts growth, with or without the presence of nitrogen. In this case, we might say that moisture *modulates* or *moderates* the effects of nitrogen. (Conversely, a healthy immune system moderates the effects of contagious diseases by minimizing or mitigating their effects.) One can also make this point another way around by emphasizing the role of nitrogen: nitrogen *amplifies* the effects of moisture. What these statements are illuminating is this: the roles of nitrogen and moisture in promotion of plant growth are *interdependent* but asymmetrical. Asymmetricality is the key feature of a different sort of model entirely—and it makes modulation cases hard to handle in a causal model. Before turning to this subject, let's examine the relationship between preemption and modulation.

Modulation and amplification involve various features of preemption. Like the first assassin in the first example, water's presence, according to our hypothetical case, is a modulator because it turns *on* (alternatively it could have turned *off*, or in some other way modified or intermediated—as does the immune system in relation to disease) the growth-promoting effects of nitrogen; likewise assassin 1 has the effect of turning *off* the effects of assassin 2. Conversely, nitrogen in our hypothetical case has an "amplificatory" effect on water's growth-promoting effects. Like assassin 2, nitrogen enhances water's effects, much as assassin 2 is poised to advance assassin 1's goal should assassin 1's efforts miss the mark. Hence, the functioning of assassin 1 and assassin 2, in production of their E, like water and nitrogen, are interlocked in a scheme of asymmetrical dependence relations.

Typically, a certain "effect" is related to the numerous contributory base "factors" in numerous such asymmetrical relations. I propose to speak of such asymmetrical dependence networks as protocols. (This will—and should—resonate with "engineering" concerns.) For a given *effect* E, a complete analysis of its contributory *causal factors* is a *filter*. By "complete" here I mean "complete for the purposes of an autonomous science." (I do not mean to give the impression of having a full story on this topic, but am only gesturing at one here, as a preliminary to a fuller story.) For any two effects E and F, their filters may overlap partially or not at all. In general, there is no a priori reason to expect that, given two effects, their filters will be nonoverlapping. This statement affirms nonreductivism, to which this entire book is devoted. To explain how it differs from competitors' views will be a chief aim of the remainder of this chapter.

The first step is to show how the asymmetricality of these dependence relations is untreatable within the cutting-edge probabilistic models of causal analysis—*path analysis*, as it is called. This is good news, because the asymmetricality of these dependence relations is a signal that there are *additional* dependence relations, and these are the ones we should be trying to capture in our scientific models. We must surrender the conception of causality to the applied sciences and capture and sharpen the other—the asymmetrical conception of dependence—for the sciences.

Not without pangs of guilt do I make this proposal. After all, following my prescriptions will cause a rift between those who utilize and maintain the concept of efficient causation and those who utilize and maintain different conceptions of the dependence relations between events and features in the world—thus rending asunder what Aristotle brought together in the name of systematics. The name of "causation," if my proposal is accepted, would be better allied with industry and the arts than with the theoretical sciences. But it is not without good reasons that I urge this—reasons that I think Aristotle would have approved. As I am arguing throughout this book, this move is in service of overall better systematic philosophy. And consequently it is in service of Science itself.

4. Why Causation Cannot Be a Lingua Franca

Cutting-edge causal modeling methodology—causal path analysis, utilizing analysis of variance—relies upon a certain paradigm for diagnosing the causal bearing of one factor upon another. The result of such an analysis is to divide up the variance and attribute portions of it (of potentially different sizes) to independent players in the causal drama (see Pearl 1988 and Glymour et al.1992). According to the paradigm, causal bearing is revealed in (as Christopher Hitchcock refers to it) "test situations": "A test situation is a conjunction of factors. When such a conjunction of factors is conditioned on, those factors are said to be 'held fixed.' To specify what the test situations will be, then, we must specify what factors are to be held fixed" (Hitchcock 2007).

In our hypothetical case, nitrogen (N), water (W), and plant growth (G) are the causal factors in question. To diagnose the causal bearing of N on G, we have to hold W "fixed," but of course since water modulates the effects of nitrogen, holding water fixed can mislead us as to the bearing of N. If we hold it fixed too low, we will not see a proper shift in the average because then we'll see no effect whatever. We will thus not be able to diagnose a straightforwardly positive causal bearing of N on G. (Effectively, rather than shifting a randomly collected sample population from one bell curve of growth distributions to another bell curve, manipulations of nitrogen will produce either no shift at all or a non-bell curve, indicating a more complex relation among the factors than causal-bearing analysis can handle well.)

To treat this sort of example, causal modeling adopts an "interaction" paradigm, on which the factors which have a bearing on plant growth (in this case two) are multiplied, so that not only do we have water and nitrogen playing independent parts in the causal drama, but also their "interaction"—a proportion of their "cross product" (the cross product of their vector representations in a certain mathematical model). This fictitious entity is modeled as an independent causal player. How can multiplying the number of factors possibly improve matters? To answer this question, we need to look more closely at the founding principles of path analysis.

Path analysis is built upon a number of important assumptions. One key example, which will play a role in our treatment of the matter a little later, is the principle of proportionate effect. A cause has a *proportionate* effect, which has as a corollary that an "increase" in the cause must bring about a proportionate increase in the effect. Another central assumption that we shall discuss now is the so-called "Markov condition," which is a generalization of Hans Reichenbach's common cause principle. Very roughly, the Markov condition stipulates that the "causes" of X must "*screen X off*" from all other variables except for its own effects, so that correlations between X and other variables disappear once conditioned upon its causes. (Less roughly, if Q_i and Q_j are correlated and Q_i is not a cause of Q_j, and Q_j is not a cause of Q_i, then there are common causes of Q_i and Q_j in the set $\{Q_1, \ldots, Q_n\}$ such that Q_i and Q_j are independent conditional upon these common causes. The correlations between initially correlated Q_i and Q_j disappear once the true cause is held fixed.) And this condition stipulates that *factors that have not (yet) interacted must be uncorrelated*. Even if this apparently innocuous assumption is true simply as a matter of fact (rather than a matter of necessity), what place does such a principle play in causal analysis? Let's return to our example.

Applying the Markov condition to our nitrogen example, we have to say that the causal factors of plant growth—which for simplicity we're postulating are three in number: nitrogen, water, and their interaction (water*nitrogen or

W⊗*N*)—must be "initially" uncorrelated. In other words, the factor *W*⊗*N* is not only an independent causal factor, but also uncorrelated with water (as such) and nitrogen (as such). Another way to say this is that we have to *orthogonalize* the contributions of each of these players. (We therefore use the Markov condition as a *regulative ideal* rather than as a simple matter-of-fact assumption. We do this in order to maintain the fiction that it is possible to attribute proper credit for a portion of variance to each of the independent causal players.)

Now, even if it makes some sort of sense to say that the interaction between nitrogen and water (the element *W*⊗*N*) is itself independent of both nitrogen (as such) and water (as such)—and indeed it does make sense because we simply construct the *W*⊗*N* factor (the cross product) so as to ensure it obeys our definition of independence—how do we construe the proposal that this interaction factor is a *third* factor? Presumably, all we are trying to get at, on this causal modeling proposal, is that *W*⊗*N* is a third thing that one needs to keep track of in order to make sense of what's going on.

When we do that, the result is a structure like the one shown in figure 4.2.

It's understandable why we'd wish to simplify: simpler problems admit of solutions via known methods—and in the final analysis only problems of a relative simplicity admit of solution at all with such intellectual resources as we might ever be possessed of. We are told time and again that only linear systems of structural equations admit of explicit solutions. So if we should posit structural equations of a nonlinear sort, we cannot hope for explicit solution—in other words, we cannot hope for a solution that applies to every set of "initial settings" on the structure. To achieve the ambitions of an explicit and perfectly general solution, we require a linear model of the structure; thus path analysis seeks to simplify in order to achieve the goal of a solution. Every textbook in path analysis acknowledges the limitations of this restriction. But what exactly does the simplification forgo? What exactly does it miss out on?

Some thinkers have suggested that the notion that causal factors must be thought of as independent keeps us from making sense of the notion of modulation.[3] I agree, and will now explain how it does so. But my conclusion will be neither that we must dispense with the project of articulating causal structures via path analysis, nor that we must somehow figure out how to improve upon it. My strategy is instead to illuminate what is missing, to say what sort of theory is required to capture it, to locate that role elsewhere (somewhere apart from a theory of causation), and finally to say what causal theory does well.

First: path analysis misses out on modulation—and other control notions, as I will now also discuss. The function of modulation is something that engineers

[3] Much of what Cartwright says in *Nature's Capacities and their Measurement* Cartwright 1989 can be read this way. See also Cartwright 2006a and 2006b.

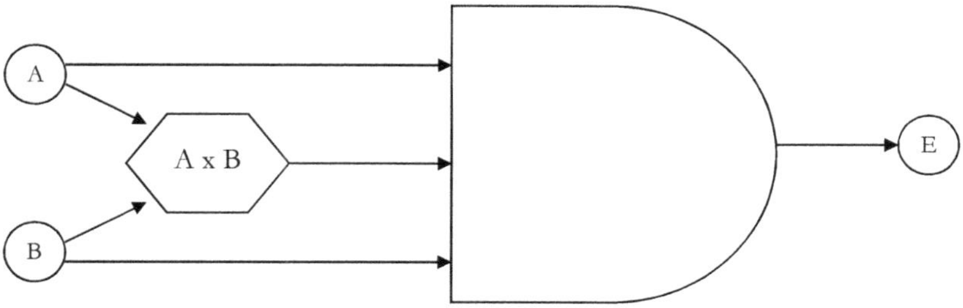

Figure 4.2 Linearized circuit

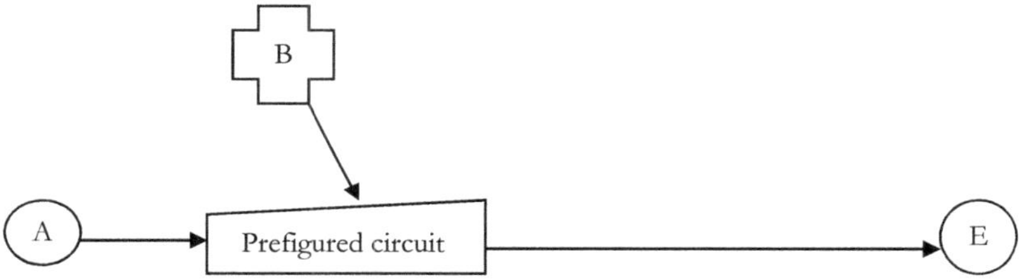

Figure 4.3 Modulation circuit. The element in the box has an antecedent functional role that gets occupied by B.

often aspire to achieve. Suppose you wish to install a light fixture that could be made to illuminate a little or a lot. What you would like to do is to "overlay" a basic electrical circuit with a kind of dimming operation that amplifies or dampens the function of the whole circuit, depending on the setting of a switch. The dimmer switch would ultimately control the amplifying function. It would control the modulator. In our original case, N has the kind of role that this dimmer switch plays. A catalyst for a chemical process is another type of modulator. Depending on the function to be modulated, the modulator can be designed in any number of ways. But its essential feature is to "intercede" or intervene between a process and and its final issue. I will represent it as shown in figure 4.3.

In figure 4.3, the role played by B is no "generic" causal role. In the structure of figure 4.3, A enters a playing field that is already prepared or laid out, contingent upon the role B plays in it. Assuming that B will enter in just like it "always does" misrepresents the role that B plays. (It would be like saying: "The switch on this box plays the role that switches always play.") The "circuit" is not just a linear circuit that simply provides entry to all factors on the same basis. Elements that fit the "box" currently occupied by B will produce a modulatory effect. The fact that this box is an integral element of the circuit should lead us to be wary of drawing conclusions about what happens when there is no modulation; that too has to be specified explicitly, where by contrast figure 4.2 tells us that A plays its independent causal "part" without regard for what happens in the other nodes. If there are circuits like the one in figure 4.3, then the world does not always present a level playing field to all entering players, as we will discuss at some length below. That, by contrast, is the sort of picture we get in figure 4.2. But if there are circuits in the world like that in figure 4.3, then we cannot always assume that the same elements will always play the same parts in any given drama. The presence of "prefiguring" elements can change an element's overall causal profile.

What we have been describing, using circuits like that in figure 4.3, is a system of representing systems that takes seriously the role of the "context" in which a factor is plugged. It takes seriously the prefigured environment for any given role. And it takes seriously the role that context plays *in making the element in question play the role it does*. A linear model does not do that and explicitly seeks to fit all the features played by a given factor in a given context *into the box dedicated to that factor*. But unfortunately a linear model is very poor at handling the representation of a species of modulators—the switches, the role played by assassin 1 in the shooting case we discussed—for the reason that a switch, by definition, is an all-or-nothing modulator. But path analysis is founded on the principle of proportionate effect: more of the cause should produce more of the effect. But a switch is not a proportionate principle: its job is to produce all or nothing of the effect associated with it. There is nothing proportionate about that. So we require the "patch" of third factors that we discussed above. Other modulators require patches as well. Patches are called in to "linearize"—and that means to make all the same kind of causal factor out of different kinds.

Of course a critic might reply that our examples are only directing attention to the fact that the approximating treatment of a system that is described by figure 4.3, by a structure like that depicted in figure 4.2, had better do a good job of picking the third element for entering in the linearized model. This is a reasonable requirement. Or perhaps, as another critic might suggest, it serves to draw attention to the fact that more elements still have to be entered—and so more ways of gesturing at the structure drawn in figure 4.3 to render a better approximation still. But no mere multiplication of factors entered into the drama will "reduce" the structure of figure 4.3 to that of figure 4.2. They simply don't share the right shape, however well the one can approximate the other locally. (This is just a geometrical rendition of the mathematical fact that a Taylor expansion of one functional form by a linear sum of simpler functions we know how to handle can only be a true equality in the limit as the number of linear terms goes to infinity.)

For our purposes, the philosophical message is this: linear modeling has to forgo treatment of prefigured functions and must treat them instead in a one-size-fits-all fashion. It has to fit square pegs, rectangular pegs, hexagonal pegs, and ovular pegs into circular holes. In other words, it has to treat the causes more coarsely than we know that they are. What is happening in the modulation cases is that an environmental variable is *coupling* with the variable whose magnitudes we are manipulating, in a particular way defined by the context. (I tried to capture this idea by giving a shape to the modulatory elements.) This immediately obviates analysis of the whole business in terms of "independent contributions" of different causal pathways. Thus the idea that the effect is under the control of the so-called "causal" variable whose bearing we are querying is inapt.

The upshot is this: we cannot truly render sensible the idea that water *modulates* the effect of nitrogen, or that assassin 1 switches on the effect of assassin 2, simply by saying that N and W (or assassin 1 and assassin 2) are each independent causal factors, as also is their "interaction." We simply wish it away with this loquacious substitution. And we miss out on the truly engineering-worthy features of switches and other modulators. The language of causal factors just masks over, for expediency's sake, the functional in figure 4.3. Still, it hasn't left us entirely empty-handed, a topic to which we now turn full attention.

Figure 4.2 is a model according to which W, N, and a link between them—all of them in some sense independent—have consequences for G, and we can make it give us an approximation or coarse description of those consequences. More importantly, it offers a way of *compounding* these consequences with other consequences, again coarsely described. This is of some value. But among other things, this analysis misses out on the asymmetricality in that dependence relation depicted in figure 4.3. (The simple statement "W modulates the effects of N" speaks clearly itself of an asymmetricality.) Eliminating the asymmetricality is fine if we are not concerned with the specifics of the relations between W, N, and G. For instance, if we're only interested in predicting what will happen in a single instance of their co-occurrence. But this gives us an account of the matter between W and G only in its broad or global features—an account of their contributions to a more coarse-grained picture of relevant factors. This coarse-grained "big picture" is a useful tool, but it's not a faithful depiction of dependence between features. It satisfies a certain set of desiderata, but it doesn't give us everything we could hope for.

There's nothing wrong with coarse-grained pictures, as long as we don't try passing them off as something else. And what we must recognize about this picture is the manner of its coarseness—that it actually does mask the fact that there is asymmetricality in the relationship between N and W. I will discuss at some length later in this chapter what we get from analysis of causation. In the interval I wish to discuss the sort of analysis that will capture what it misses out on.

What do we require in order to render faithfully modulation (and related structures) is a notion of *control*. As will be explained, this comes only in a Systems analysis.

Thus one of the fundamental signals of this chapter is that there are several directions one can go once one recognizes the failure of a path analysis to capture the structure of figure 4.3. Each of these corresponds to a different goal or aspiration. The first direction, and the one chosen by path analysts, is towards a "big picture" perspective that blurs details of modulation and highlights instead a rough contributional "what has bearing on what." This is what we mean by a causal picture. It is no shame to be involved with the production of such a picture. But it must not be confused with an enterprise in which the details of figure 4.3, and others structures still, are given a faithful analysis. That would be a scientific enterprise. Nor must the former be confused with the analysis we are engaged in assembling now in this book—an analysis of the types of dependency relations into which categories of quantities may enter.

This second aspiration is also clearly not a scientific enterprise proper: it is instead a more general picture—thus a metaphysical picture proper. But it is not a causal picture. It is instead a picture of the nature of such things as modulation. More generally, it is an analysis of the variety of dependencies, their nature, their origins, and the general metaphysical structures and hierarchies that they create or enter into. In this space belongs analysis of how Systems, as granules, hold together. This is a topic I shall focus on at some further length in what remains of this chapter before moving on to analysis of how spheres of specialized scientific content are created.

Before we move to the topic of Systems however, I want to be clear about my reasons for discussing the assassins case—and modulation cases in general. Yes, it is to emphasize that they cause trouble for linearizing models of causation (and so quite importantly, for the path analysis and other probabilistic models). But of course I'm not the first person to point this out—the problems have been familiar to many researchers, who have collaborated on a very considerable body of interesting research on the topic. The reason for bringing these cases to our attention here is to offer my diagnosis of why they pose a problem. They pose a problem, as I've been at pains to explain, because the models aim to force all causation to fit the model of a directed acyclical graph—figures like 4.2 and 4.3, but with only one sort of shape for causal factors, or (equivalently) figures without shapes but only nodes and arrows. But a true description of the operation of a switch (for instance) requires more nuance than simply the language of "positive correlation"; and in particular, a switch's all-or-nothing effect requires remarking upon—in those terms. (Notice that I have made no commitments, in this paragraph, as to whether there really exist such things as true switches, which play that causal role in real life. Because the points stand, whether true switches—which are the hardest to represent by a linear model—exist or not.)

The point of the examples is therefore to say that we, as intelligent scientific theorizers, have no trouble understanding the operation of a switch. But that linear modeling seeks to "turn off" that understanding, to minimize it. And this is a mistake. Scientific understanding cannot do without switches, even if there were no actual switches in the actual world. The switch model might be fundamental to scientific modeling, even so. We will need eventually to bring switches back into the fold of science. Beginning with the next chapter, we will seek to make room for them.

5. Control

To lay the groundwork for the analysis of dependency relations, which starts to take shape in chapter 6, we will begin with a very rough analysis of the notion of modulation. It is not my intention here to achieve a decisive analysis so much as to gesture at the sort of thing needed. To work out the specifics merely gestured at in this preliminary analysis, and from there to launch a study of complex dependency relations and their interrelations to one another within the taxonomy of such relations, goes well beyond the scope of this book, which seeks only to urge upon researchers the need for studies of this topic. In any case, my preliminary analysis of modulation begins with the notion of *control*.

To control is *not* simply to stand in a causal relationship (by that I mean some kind of probability-relevant relationship, as presumed in path analysis) to some event, or even to a network of such events. Control is a matter of *authorities*, a matter of which directives have a trumping effect (winning out when there is conflict, and herein lies the autonomy of the filter). Control is not essentially a local matter of where a process originates, nor is it a matter of which features of the world ultimately explain what the products of the process ultimately look like. Control is a global, all-over affair, a matter of *rank* and *protocol*, and not well correlated with the outcome or goal of a process or procedure.[4] So it is very difficult to diagnose when encountered on the ground.

[4] This question is directly related to the question of whether a process is *centralized* or *distributed*, and this is a broadly logical or computational feature of the process—it pertains to a design attribute, rather than to features special to physical implementation. A centralized process is a hierarchical (or top-down) process, with a single "processor" residing at the topmost level of control, whereas a local or distributed process is one in which control is distributed horizontally at the topmost level among a plurality of processors equally ranked from the point of view of control. We might say of distributed processing that the initiative is "grassroots." Obviously control can, at least logically speaking, be distributed at intermediate levels as well; a process can be centralized at the topmost level but distributed at lower levels, and vice versa. Norman and Shallice (1986) present an intersection model of top-down and bottom-up processing, without a ranking central processor. Hardcastle (1995) makes a beginning for a philosophical discussion of these topics. A fuller account of the difference between control structures and causal structures is discussed in Thalos (2007).

Control is best illustrated by examples where by hypothesis we know what the *protocol* is or simply where we stipulate the control structure. Military organizations or militias are perhaps the best illustrations of what we may call *hierarchical* control, and other examples of control are contrasts to it. Protocol is not concerned so much with what forces, factors, or influences are actively exercised (changing or being changed) during the relevant real-time process, so much as it concerns a global feature of the organization (as well, potentially, as accountability practices) involved in that process. Control over my car, for example, is what *I*, as driver, lose when the brakes of my car fail at the top of the hill, even if I have occasion to call upon their service (and then entirely without success when I do) only at the bottom. During the period of time when I have no occasion to call on their service, matters will proceed exactly as they might well have done had I indeed enjoyed control over them; still, I do not enjoy that control. Elsewhere I have adopted the following rough and provisional analysis of *authority* or *rank*, preliminary to a mature account:

> *Axiom a.* An entity, unit, or function A *ranks over* a second entity, unit, or function B when the edicts or instructions or processes of A win out over those of B in cases where the edicts or instructions or processes of the two are in conflict or cannot go forward simultaneously.

Now, this axiom is palpably rough, and its roughness is instructive. Nothing in it takes into account any of three important qualifying ideas that will ultimately have to be handled in a mature conceptual treatment of control: (1) ranking can be complete or partial, in other words, that it may leave entities uncompared as to rank (*partiality*); (2) it may be circumscribed rather than global, applying to a very clearly delimited range of operations (*circumscription*); and (3) these two ways that a rank can be qualified or circumscribed might intersect or overlap in such ways as quickly confound diagnosis of the precise balance of partiality and circumscription on the ground (*tempering*). But ultimately, these three ideas will be important for articulating the notion of an *autonomous* filter for an *autonomous* science. For now all we need is the simple conception of ranking, as a piece of the relevant metaphysics for our examination of Systems analysis: it allows us to conceive of control structures as structures that overlay pure stochasticity or preexisting dependencies and govern when and where these "controlled" structures interact. The new arrivals choreograph and direct. They are the stuff of upper management.

One concept that has been common in the disciplines of psychology and neurophysiology since the 1950s is that of *inhibition*.[5] This is indeed a genuine

[5] It originated in the 1950s in the work of Eliot Stellar, in the form of dual-center theory, now generally thought overly simple.

control concept, much more than it is a causal one. The concept of inhibition is (roughly) the concept of veto or brake—of one-sided modulation (very much like the absence of water in our hypothetical gardening example). The controlling entity or process prevents an action, even one that is already going forward, from going to completion. This is roughly one-half the concept of ranking. The full concept also includes the controlling entity being in a position to "arouse" or "excite" as well as inhibit. (This idea too has been in circulation in physiology and psychology, in the theories of memory, attention, motivation, and learning.) Put together, arousal and inhibition would seem to be two necessary, and perhaps also jointly sufficient, pieces of the concept of rank. Perhaps it will turn out that to achieve more conceptual traction with the idea of ranking, one will need to break it down further into these two component ideas. And it seems at least initially plausible that the two halves of the control function can be separated—that one entity or process can have inhibitory control over a second, but not excitatory control. It would serve us to investigate these sorts of questions further, at least conceptually. When once a full complement of the necessary control concepts have been articulated, it will be possible to give a detailed account of homeostasis and similar physiological and systems notions in purely metaphysical terms.[6] For our purposes here, Axiom α will have to suffice.

It's pretty clear how the idea of ranking handles modulation, at least in relation to the nitrogen and plant growth case: water's absence plays the role of a veto in that case, no matter how much else of what's needed for that process to go forward is present. But this is only one sort of case of modulation. There may be many more types. It is not within the scope of this work to work out the details of the nature of modulation. I point out water's role as a modulator of this type to suggest that this one realization of the modulatory function is probably key to understanding the origins of filters as wholes—of systems of order as such. This is not the place to develop this idea, but only to suggest the importance of the concept of modulation for the emergence of order.

In light of these ideas, my position is this: the fact that causal language is inapt for handling analysis of modulation is a fundamental reason for being skeptical that causal language can be the ecumenical language of science. Preemption cases are thus cautionary tales that should direct us away from simply taking as unproblematic the proposal that causal notions (and the intuitions they would systematize) can be harmonized in a way that renders the central concept of causation a lingua franca of all scientific communication.[7] My contention is thus

[6] The metaphysics of systems theory is currently in its infancy. Thalos (2007) makes some progress in articulating the metaphysical differences between systems theory and causal analysis.

[7] Many of those who labor to craft consistent philosophical accounts of causation have high aspirations: they aspire to tame these relations (the precise term is "reduce" them—to reduce further the modest amount of complexity in the circle of causal concepts) to their favorite relations—counterfactual

that the language of causation cannot handle the burdens of the aspiration. Still, without aspiring to "translate" all the findings of science, causal concepts might nonetheless help to produce a "broad-brush" picture of the world in nonscientific language. It helps to produce an "aggregate" picture, but not an altogether correct one. Here is what I mean.

How, given the difficulties we have encountered to this point, should we think about the cumulative effects of all the relations manifest in the world among the various local features? What makes the cumulative picture difficult is the fact that there will be "overlaps" in which some "region" is independently governed by *different* Systems of organization, where some effects of an autonomous filter are "dampened" and others "amplified" by a superseding or controlling filter. The best we can do by way of managing the cumulative effects of these laws in a comprehensive fashion is to talk about probabilities and probability-changing relations, but decidedly *not* to suppose that they are guides to some more lofty and comprehensive systemization. Probabilities (which is what path analysis is all about) are just rough guides to action for us who require some sort of guidance, and nothing more. They are "best guesses" or "best estimates" of the rates of event occurrence in light of all that we indeed can know about the multiple domains, in the ecumenical language of probabilities. That's all. We cannot simply expect an accumulation of such guesses to amount to an aggregate body of knowledge, but simply a way of keeping books in a rough language to help keep cumulative track of useful (but of necessity incomplete) coarse-grained features of things that bear upon a given event frequency. Not nothing. But not everything we can hope for either.

In fact when we look at the special sciences on their own, none of them seems to be concerned that their language can harmonize with anything like a lingua franca of sciences. They go about their business as usual, unconcerned about troubles with preemption-type problems in the theory of causation. Naturally, individual scientists use the language of causation for ordinary human transactions, or as loose language when in casual conversation with colleagues about their research. (After all, it is the language of common sense.) But causal language is quickly displaced when sleeves are rolled up, and return is wrought to the official language, often formal or mathematical, just as the official tongue is the one utilized to report findings in journals.

But if causal language does not provide a language for the treatment of what goes on within filters or what transpires between filters, what language will do

---

relations or probability-changing relations—so as to be able to say that causal relations are configurations of these favored relations (and consequently that causation is not mysteriously noumenal or superadditional to these more "basic" relations). The ultimate goal, even if no one theorist has put it quite this way, seems to be that causal language is the common currency by which all the sciences can trade their wares, as well as offer them for sale more broadly in the marketplace of ideas.

this? My answer is Systems analysis, the ecumenical language of engineers. This answer effectively liberates the engineer from her formerly humble location, within the applied sciences, and elevates her status to that of a philosopher—well, at least the engineer concerned with the theoretical aspects of the notion of engineering as such. This elevation will occur in chapters 6 and 7. In the meantime, we will take up briefly an analysis of what causation might be able to achieve.

6. What Causation Is Good For

Let's focus for a moment on the conception of causation that dominates today in science, especially biological and social science today. This conception is reductive in the following sense: it maintains that causal relations are nothing over and above statistical relations (relations of covariation among variables and of co-occurrences among events). I want to take a moment now to talk about reduction of causation. While this conception of causation is nowadays the only one encountered in the social sciences—the sciences in which the notion of causation has been fought over—originally it had competitor conceptions. And it is only recently that these other research programs in analysis of causation have grown largely inactive.

I have in the past argued in favor of the reduction that now prevails among scientists, especially social scientists, for many reasons. This project has its roots in the work of Hans Reichenbach (see Thalos 2002b, 2003), who argued that the central principle around which the reduction now proceeds (the common cause principle) is a founding principle of the Theory of Relativity—which is our best theory of the universe at large scales. Furthermore, the reductionist conception is the only one that converges with the most robust, coherent, and well-studied (as well as most-extended) axiomatization of causation—that of Judea Pearl (first appearing in Pearl 1988). I have also argued that the reductionist conception explains our puzzlement regarding some of the more exotic correlations in the quantum regime—because these correlations violate the common cause principle (and therefore Pearl's axioms). So giving up the common cause principle as encapsulating the idea of causation—giving up the probabilistic reduction—would render these correlations completely unproblematic. And in so doing, we would fail to respect an important line that the human intellect draws.

Before proceeding I want to make it clear that I still stand behind these arguments. I still think that the reduction is the most coherent analysis of the concept of causation—a concept other animal species share with us, at least operationally. They would find quantum-mechanical correlations puzzling if they could understand them. But though I stand behind the reduction, this doesn't mean that I endorse causal analysis as the quintessence of scientific reasoning and theorizing. This will be the conclusion of the current chapter. Causal reasoning might comprise some component of scientific theorizing in some areas of science. I don't wish to rule that out (though I am inclined to think that whatever that component might be, it's got to be pretty small, and pertinent primarily in practical or clinical applications of the sciences in question). What I do think is that causal analysis is best suited to thinking about how to make things happen, at least in the most global way. It's the way to think about making things happen when you don't have the engineer's story about how the system is put together from scratch.

I also want to insist—and this is the cardinal idea in this chapter—that there is more to scientific reasoning than this.

7. The Reduction of Causation

Ever since Hume published his provocative thoughts on causation, there has been enormous controversy over how, philosophically, to handle our very conception of it. Hume himself took for granted that philosophical handling of those things that cannot be directly observed (like, for example, the relation of cause and effect) is quite different from handling such things as leave a trace or signature upon the senses (like, for example, ordinary objects and their perceptible qualities).

But some of Hume's successors were of a different mind. Immanuel Kant and Edmund Husserl, for example, on opposite ends of a certain epistemological spectrum, nonetheless agreed between themselves, but disagreed with Hume, on the question of whether the unobservable should be handled differently from the directly observable. They understood all skeptical challenges as continuous with Cartesian doubts, making no discriminations of kind among varieties of skepticism. Thus their respective treatments of skepticism have nothing special to say about cause and effect as such. When they give an account of our knowledge of cause and effect, this is in no interesting sense different from their account of our knowledge of ordinary workaday objects. This failure to differentiate is in my view ill-advised. For the category of cause and effect is substantively different from—among other things—the categories of object and observable quality. It demands special attention, for the very simple reason that we don't initially know what we're looking at, from the standpoint of metaphysics, when we are contemplating the relation of cause and effect. The category of relation under which the relation of cause and effect falls is not a given. For is it given how the relation of cause and effect differs from the relation of larger-than? From the relation of equidistant-from-point-A? Since the answer is no, we cannot be certain that the treatment of skepticism that befits the case of everyday objects, or even persons, will apply to the case of cause-effect equally well.

The point needing emphasis here is not the obvious one—that knowledge of the indirectly known is indirect and hence more difficult, complex, or (if you prefer) impossible. That's really the point I wish to *resist* making. For making that point rests on having already accepted (groundlessly) the proposition that, as regards metaphysics anyway, the category of cause and effect is in no significant way different from the category of object—that cause-and-effect relations are just as out-in-the-world, equally subject to evidence, equally subject to being known. But this might be false. It might be that the reason we do not have direct evidence of cause and effect is deeper: that cause-and-effect relations are different in kind from the relations of larger-than or equidistant-from. Handling the epistemology of cause and effect must therefore begin with a treatment of its metaphysics. In this regard, philosophical treatment of causality might be importantly different from the treatment of other relations. Philosophical handling of causation is in substantive ways more like philosophical treatment of theological or moral matters, in its degree of theoretical foundationality. Handling causation philosophically requires the same sort of attention as handling knowledge (and more specifically, science) itself.

In the recent past there prevailed three quite distinct understandings of the enterprise of elucidating causation. The first and most visible is what may be termed the *semantic* understanding of the enterprise and is preeminently exemplified by J. L. Mackie's highly influential book *The Cement of the Universe* (1980) (variants, but in the same school, are the accounts of Lewis 1979 and Mellor 1988). The semantic approach consists first and foremost in seeking a conceptual analysis of the term "cause" and its cognates in everyday life. The second, and what I shall refer to as the *definitionalist* understanding of the enterprise of elucidating causation, looks to Reichenbach as its hero. This approach seeks an elucidation of causation via an examination of the so-called fundamental physical theories of the universe. I call it *definitionalist* for the simple reason that on this approach Physics, by *definition,* is engaged in the business of illuminating causes and in such a way as ranks it as a higher authority on the subject of causation than, for example, Biology.

The semantic understanding presupposes a certain uniformity in our everyday uses of the term "cause"—namely, that there is one unique, or at least one core, causal concept, or at any rate one primary focal point in meaning of everyday applications of the term. Real-life deviations from this ostensibly singular usage, once having come to light, are typically legislated away as either metaphorical or peripheral usages, or even as downright misuses of language. The definitionalist understanding, on the other hand, presupposes that there is a certain fundamental relation among events, that this fundamental relation is what is meant by "cause," and that (furthermore) it is causal relations that are illuminated by fundamental Physics. Evidence to the contrary of these presuppositions is never a matter of concern, less yet of discussion.

The third understanding—which gives rise to the *reductionist* enterprise (which might equally well have been called *revisionist*)—is subtly different from its two compatriots. This understanding resides in the reductionist's two-step project. First, the reductionist seeks to identify a core conception of cause, as the central function of the causal concept in the economy of human life, without necessarily seeking to identify it as a focal point of meaning, or even as a focal point of scientific meaning. For in the eyes of the reductionist, the fact that in everyday life, or even in a given scientific enterprise, some or even all folks are disposed to accept or reject some statement couched in causal language, under specified conditions, is fundamentally insignificant, in itself, for the elucidation of causation. Neither does the reductionist seek corroboration of the central conception in fundamental physical theories or physical phenomena, for the reductionist need not assume that Physics aspires to describe the world in causal terms. In fact, the reductionist, as reductionist, takes no position of any kind on the physicist's professional occupation.

One of the most resilient reductionist proposals—and indeed the one that prevails today—can be traced back to Hume himself. It consists of the idea that causal facts are nothing more than configurations of certain facts about chances—specifically, facts about certain factors enhancing or diminishing the chances of others. This is the proposal that lies behind all probabilistic accounts of causality.

After identifying a core conception of cause as the crystallizing notion around which causal language can be understood functionally, the reductionist subsequently seeks to demonstrate that all potential propositions, true or not, that are worth expressing in causal language translate comfortably into language openly referring only to facts of the preferred kind—for example, facts about chances or chance-raising. (For example, the proposition that smoking causes heart disease comes out true, and acceptable, only on condition that so-and-so facts hold about objective chances for each sort of event.) This, as the reductionist sees it, is a proposal concerning how to proceed with precisifying common parlance for the purposes of a particular discipline such as medicine or engineering. The reductionist then proposes to explain this remarkable fact, that causal formulas can be successfully and entirely replaced by formulas in a preferred language with well-articulated rules of inference, by the *metaphysical* proposal that there is nothing more to causal facts—at least for the purposes of these disciplines concerned with precise formulation of them—than facts of the preferred kind. If there's nothing more to causal facts than this for the purposes of engineering and medicine, then there is nothing more to causal facts, unconditionally.

The reductionist approach to causation, unlike the definitionalist approach, does not presuppose that there is a unique fundamental (causal) relation among events, or that the most fundamental physics undertakes the illumination of that relation. It claims only that, where there is a causal fact, this is nothing over and above certain facts of the preferred kind.

The point is simply that the reductionist is operating at the level of conceptions and their economy in human life. To put the same point somewhat differently, the reductionist is concerned purely with the *metaphysics* of causal facts as such, not with the empirical question whether the world we live in is a causal one (a question that the friend of the definitionalist approach, true to name, answers definitionally).

Thus the reductionist is decidedly *not* concerned, *as* elucidator of the conception of causation, with whether particular events in the world conform to any imperatives of causal regularity. So it serves the antireductionist badly, for example, to criticize the reductionist by drawing attention to the fact that the reductionist account does not accommodate quantum correlations.[8]

But why favor such a construal of causation? The best reason (as Reichenbach 1956, 1958 realized) is the reason Einstein was disposed to give: that EPR-type phenomena (exemplified in the quantum correlations we encountered, for example, in chapter 1) violate the relativistic principle that no one thing can act on another at a distance without going through intermediaries—that, effectively, there can be no causal action at a distance. This particular case proves—decisively, to my mind—that we cannot simply be definitional about causation; that is, we simply cannot, definitionally, view Physics as illuminating causal relations. A certain precisification of our ideas vis-à-vis taxonomies of event patterns indicates clearly, and in advance of observation, the sorts of correlations that can and cannot be subsumed under causal patterns. The common cause principle dictates that certain patterns are inadmissible as causal patterns, and a taxonomy that fails to take notice of this line (between the patterns that are admissible as causal and the patterns that are not) would fail us miserably.

Now it does not follow from this datum alone that we must be reductionist about causation. But it does show that there is considerably more at stake for us—as denizens of the world—in the concept of causation. It implies that even before we do Physics, we are dividing between the patterns of events that can and those that cannot be subsumed under a causal understanding of patterns. This datum speaks quite decisively against the definitionalist answer to the question of elucidating causation.

By now it has become apparent that this entire chapter has been a plea against a definitionalist answer to the question of elucidating causation. More than that,

[8] Numerous experiments that demonstrate quantum entanglements were inspired by Einstein, Podolsky, and Rosen's criticism of quantum mechanics, now known as the EPR. John S. Bell formulated a theorem (this is the Bell's Theorem referred to in chapter 1) to the effect that EPR-type findings are inconsistent with a local conception of action. For more on these matters, see the numerous essays in Cushing and McMullin (1989).

I have been at pains to lay out reasons for why scientists should surrender the notion *entirely* to those who have better uses for it. This will serve Science better—as the rest of the book is devoted to showing. So, against the definitionalist attitude toward elucidation of the notion of causation, I would say that our philosophical account of causation should leave it open as a genuine possibility that our world does *not* conform to causal expectations, whatever these happen to be. For it should be possible that our practical concepts (as contrasted with those concepts that serve purely scientific goals)—the cognitive instruments that evolution has managed to instill in us—are incapable of application to all patterns of events in the world, if to be so capable would be more costly in the economy of human life. It should, in other words, be possible to discover that our ordinary, everyday cognitive tools are rough-hewn and not suited to absolutely every scientific endeavor—as contrasted with those endeavors that are imperative to individual human survival. It should be possible to discover that our ordinary cognitive tools serve certain uses but not others. Also it should be possible to discover that the proper aims of successful "fundamental" physical theories—which are decidedly not captive to imperatives of human survival—are not the confirmation of our causal expectations about the world. Once we discover this—and I venture to say that we have—we should consider surrendering the concept of causation to those (the clinicians, for instance) who can make good use of it.

Science, by contrast, is about (among other things) how systems of various types have come about, in other words, how they have come together from scratch. Of course "scratch" is different things in different sciences. The "scratch" of Physics is certainly not the "scratch" of Biology. In what remains of this chapter, I will discuss the differences of aspiration among the sciences. Chapters 6 and 7 will then take up the question of how the sciences as a whole deal with the question of how systems are put together from scratch. It will be a discussion of the core principles of Systems theory.

8. The Variety of Sciences

Whereas in ordinary, nonspecialist life we all care about causal generalizations and applications, general and practically useful construals of the world à la figure 4.2, the sciences are all about the specifics of what I will from here on out refer to as "filters." Figure 4.3 speaks of one element of a filter. Filters are made up of many elements like that in figure 4.3, all of which come together to comprise the subject matter of a scientific area. To the extent that each area of science is concerned with illuminating the elements of its own filter, to that same extent sciences share a "logic," a "methodology," or a "metaphysic." To the extent that the filters of the world are largely independent, the areas of scientific concern can be treated as autonomous. Finally, to the extent that there are a multitude of independent and largely autonomous filters, there can be a multitude of independent and largely autonomous sciences. That said, it might be worth exploring the question of how different sciences go about their business.

How do filters work? The answer, as I will phrase it initially for expository purposes, is that filters make an *engineering difference*—and this is where protocol comes into it. As you'll recall, every network of would-be causes (e.g., firings, as in our assassination example) in any preemption-type case might also stand under a further control protocol—an *organization* of those first-order factors that we have been referring to as their filter or *protocol*. The protocol stands over and above these would-be causes. In our example, C1 is a controlling factor in the situation, and whether it obtains determines which first-order process goes forward to bring about E. To state these facts, we required the notion of a *protocol* to govern something that stands in a relation of *lower-order process* to it. These notions are all that is required to articulate the nature of a filter.

The protocol summarizes all the facts at filter level. No further story about "causation" is wanted there, or welcome. Furthermore, no treatment of what lies "below" filter level—the controlled quantities—is wanted. True, the uncontrolled quantities exercise an impact on the output, but that is not part of the proceedings at filter level. Such facts are simply taken as given—as inputs into the filter. Likewise from the perspective of filters that intersect or overlay entirely the current one (if there are such things): the outputs of this filter function as inputs into these and are accordingly treated simply as givens. This is why it seems as though there are "barriers to causal penetration" between filter boundaries. There are in fact no such barriers, but for the purposes of describing the proceedings at filter level, the "shape" of what comes in is treated simply as given. (I will argue in the next chapter that there are no barriers.) We might pretend causal independence, since the pretense does no harm, but this amounts to no more than simply taking the results of governed or controlled quantities, including those that are outputs of other filters as given.

Protocols organize new rules under which processes of previously existing domains or principles get further organized: protocols introduce new rules into old structures. (I'll refer to the previously existing domains or principles as "older," because the "age" of an order, in relation to the entities that enter into it, is much more ontologically significant than the scale on which it operates.). Presumably the filters of Physics and Chemistry are much older than those of Biology. Thus Biology takes filters from Physics and Chemistry as "given." They are the "scratch" from which biological systems get built up. Obviously the filters of Physics and Chemistry are not *their* own "scratch." This leaves us with some questions for Physics and Chemistry—questions on which they have been admirably setting themselves to work: What are the elements of our "scratch"? Which can we take as "given," and which can we show are decomposable into the elements? How are the decomposable ones assembled? What are the laws that govern the process of their assembly?

What's more, Physics and Chemistry are closely related. The extent to which they are autonomous is still under contention. Their boundaries are continuously being renegotiated. The same can be said for many more sets of sciences: Psychology and the neurosciences, the social sciences, Ecology and the other biological sciences. The boundaries between sciences are constantly being negotiated. It is important that boundaries be negotiable. The negotiations are signs of the proper operations of science: always working to test the limits and boundaries of the filters with which they are concerned.

Here is an illustration of how filters work, which suggests how the boundaries between disciplines need negotiating. Consider two neural processes, say A and B. Process A alone would raise the arm up; process B alone would push it down. Under a strictly electrochemical regime, there is no telling what will happen when these two electrochemical processes meet on a strictly level electrochemical playing field, because the electrochemical regime does not seek to resolve such conflicts in any straightforward way. (Maybe the arm just stays in place, or maybe it will go up or down according to some differential between the two processes, if they were to meet on a strictly level playing field.) So, by the lights of electrochemistry:

> *Axiom Electrochem.* When process A results in neurons upwind of (for example) the hand sending an impulse with characteristics χ_A to neurons that attach to the hand, and process B results in other neurons sending an impulse with characteristics χ_B, the neurons of the hand will receive an impulse $\chi_{NET} = \Phi(\chi_A, \chi_B)$.

Here $\chi_{NET} = \Phi(\chi_A, \chi_B)$ is a very complicated working out of electrochemical principles.

But this is not what happens in real-life neurons, because neurons are really under the governance of a further filter—the neuronal filter, as we may refer to it for gesturing purposes—which exerts a "new" measure of control over the electrochemical processing. Neuronal processes exert an order—a ranking order—over electrochemical processes, by imposing another axiom:

> *Axiom Neuronal.* When process A results in neurons upwind of (for example) the hand sending an impulse with characteristics χ_A to neurons that attach to the hand, and process B results in other neurons sending an impulse with characteristics χ_B, control structures C will suppress one

of those impulses, and the neurons of the hand will receive an impulse $\chi_R = \chi_A \vee \chi_B$.

In engineering terms, process C will act as an *OR* gateway. A gateway is a control structure. It governs the pathways by which "older" structures interact. Note that χ_R and χ_{NET} may actually coincide (may be equal in value for all actual input values), and the facts would still be different: in one case the output is under control, while in the other the outcome emerges as a result of an uncontrolled process. This is a difference that matters to scientific analysis.

However, routinely the two (χ_R and χ_{NET}) will be quite different. Ordinarily χ_R will bear a much simpler relation to χ_A and χ_B than χ_{NET}, and so it will be quite obvious (once we understand the nature of the processes that are under their governance) that governance has been imposed: filters bring order rather than multiplying or amplifying complexity! This is a topic that will be revisited in chapter 6 when we further discuss the nature of the cement of the universe.

What is important to keep in mind as we transition to the next chapter is this: the cement of the universe is its ever-evolving set of filters. At one time the universe held no more than a set of physical elements. (Perhaps this stage was its very beginning, perhaps not.) But with time, new and more organized things came to be. And with their existence, new possibilities for new forms of organization came to exist. With each new order or organization, new avenues for further organization reached the horizon of possibility.

This is the real cement of the universe. Nowhere in the sciences is that process described in any careful or scientific way. Indeed, the very description of the process stands in tension with any number of principles in (for example) the theory of thermodynamics. And this tension requires treatment and dissolution, which cannot happen until we can treat the systems of order within a single lens of scientific inquiry. Bringing this topic into focus, as a subject of scientific inquiry, is the central goal of this book.

5

Moves and Movers

Notes on the Progress of Science

All paid employments . . . absorb and degrade the mind.
—Aristotle, *Politics*

In the last chapter we surrendered, on behalf of Science, the language of causation. Science does not require that language and is better off without it. Each scientific discipline has to devise its own (usually asymmetrical) language of dependence, since the language of causation cannot be a lingua franca. This might be construed as a cause for alarm: for if Science were to abandon the language of causation, then doesn't it abandon with it the highest standards of proof we know of, and across the board? And aren't we then cast adrift, methodologically and epistemologically, if we give up this language as the logic of Science? No. For three reasons. First, the logic of causation has never been the logic of science itself—as the remaining chapters of this book, starting with the next, will discuss at some length. Second, no methodologies are given up when we signal our acknowledgment of the fact that the language of causation is not a lingua franca for science and accordingly replace it with the language of *dependence relations*—a language that we shall begin to articulate carefully in chapter 7. It is important, as we shall see, that we stop saying "cause" when we mean (and have meant all along) "depends on." And in the meantime those who are really and truly looking for causes may proceed apace with what they do best, enjoying proprietary usage of the term. Once we stop using the language of causation haphazardly, we will attain an important clarity on the subject of how what one scientist does is related to what other scientists do. Third and finally, we actually gain rather than lose. Because then we shall have an additional metaphysical anchor: we can invoke standards of analysis related to *dependence* as such. These will provide further checks. So, at least in principle, we can attain higher standards of quality control overall than before, not lesser.

You might still be concerned with how things shake out on the ground, even if in principle things should work out for the better. For I am proposing we cede the logic of causation, in some sense, to the clinic—to the applied aspects of Science. The practitioners in the clinic, doing applied work of various sorts, will be left with the job of hunting for causes. And we—that is, theoreticians and scientists—will not be in a position to stop them calling it Science. This really does have its own kind of risk: the risks of politics. It's not for nothing that sociologists have remarked that money, which is plowed in large sums into the hunt for the causes of our ills, by both governments and corporations, corrupts and bends Science.[1]

The topic of this chapter is precisely this blowback from the clinic—conceptual innovations that are rooted in the practice of applying Science to practical contexts, but which may pose a risk of unintended harm to Science itself. Would the proposal for which I am agitating, asking Science to cede the language of causation, pose special risks to science? I shall answer no.

1. Science and Politics: The Question for Our Project

It is a temptation to suppose a natural instinct among humans to divide *entities* or *properties* into basic kinds, as a way of responding to certain preexisting facts about reality—the ones associated with natural kinds or natural universals. On a simple account of this matter, we respond to the fact of natural kinds when we categorize entities. On this simple view, natural kinds are themselves part of the basic furniture of the universe. But the reality—a reality to which the rather recent discipline of science studies, with some help from cognitive science, has drawn attention—is that our diagnostics (and this is no less true vis-à-vis causal diagnostics) are typically guided by background beliefs in general, but also the following things in particular: (1) *ideologies* on the subject being studied and (2) the *motivations* that impel us to study the matter under scientific investigation, comprising both more and less overt reasons we are seeking scientific answers in the first place. Both ideologies and motivations exert influence upon the philosophical orientations we take toward the categories of the world. Nothing can be

[1] This is not the place for a disquisition on the politics of science, but the facts are indeed quite shocking. McGarity and Wagner (2008) is a compendium of such facts; compare Oliver (2006) for discussion of some issues to do with nutrition and public health. This is the moral hazard of our times: as a society we offer smart, political animals financial incentives for innovating ideas that "solve" "problems" and hold them harmless when these innovations go wrong. The consequences are then borne by Science itself, together with other third parties.

done about ideologies: they are simply a fact of scientific life and indeed may be important for generating new ideas for scientific testing. Quite often their impact is easy to discern. Less easy to discern, and often hidden from everyone concerned, are the motivations. So the questions I wish to pose here are: what is the impact of intellectual goals—specifically, what is the impact of clinical goals held by researchers with orientations toward application contexts such as medicine and engineering? And how is that impact different from the impact of goals of investigators without such concerns? Would banishing investigators who are officially nonclinical from working with causal terms, as I am proposing we do, have a negative impact on quality control?

For the purposes of these questions, I am less concerned about the most high-profile cases—such cases, for instance, as that of climate change. While it would of course be a travesty to settle the question of whether or not there is global warming, and if there is, what is its cause, on political grounds, any errors that might be made in regard to such questions will be errors of *fact*. We will simply answer the question incorrectly. I am concerned here with the impact on the *conceptual* arcs or trajectories in a given clinical inquiry—the impact of innovations of concepts or purported natural kind terms. Errors in this regard—of which the question of global warming is innocent, whichever way it is answered—are liable to do the most damage that Science can sustain. It is conceptually easy to recover from an error of fact as to whether something is a cause of something else, less easy to recover from an error like, for instance, calling some process cancer when it insufficiently like certain paradigm cases.

We will focus on two case studies—one drawn from engineering and another from psychology and medicine. The first we've encountered before: the science of flocking and its applications to control of a fleet. The second is the case of autism. We will see that the language of dependence, which I shall be proposing as the lingua franca of science, can help us navigate the terrain of quality control better than the language of causation, because the language of causation is concerned with questions of fact, rather than questions of how concepts are related to one another.

2. Science and Politics: Two Types of Pressure

Before we proceed, we must recognize a distinction between modeling premised on purely theoretical considerations and modeling premised at least partly on something else too. I call the first of these the *theoretical* context of scientific theorizing, and the second the *practical or clinical* context. What, if anything, should be the bearing of context on what we do with the models produced within them?

First, it can be quite hard to disentangle in practice the two types of contexts. This is the hard lesson we have learned from several decades of sociology of science and science studies. Science is a chorus of unelected voices, each of which can exercise a disproportionate influence on the intellectual legacy of future generations and public policy. But here we are concerned, not so much with the influence of Science on society, as with the influence of society on Science. This latter is by no means inconsiderable.

It is important to distinguish between two types of potentially "bending" influence upon Science. Both types have to do with the effects of the motivations of scientists upon Science, and not with the effects of ideology or background assumption. First, there is the type of bias documented in McGarity and Wagner (2008) and Oliver (2006)—bending that is motivated by economic gains. This, as McGarity and Wagner put it, is *outcome-oriented* science—science whose "findings" are dictated from the start. This usually involves no conceptual innovation, but typically there is a proposal to answer a causal question of the form "What causes X?" with large policy and/or market consequences. When the outcomes to such investigations are motivated by the economic concerns of parties whose primary interest is *not* in Truth, I'll refer to these as cases of *political influence*.

Second, and potentially more harmful, is the sort of influence that is motivated by good intentions, but not necessarily with science in mind. They result from well-intentioned people trying to carry out perceived imperatives to help persons in need or to improve the conditions of human life generally: heal people's ailments, build bridges, and so on. These are obviously legitimate concerns and quite often explicit or overt. Still, in such cases science can suffer long-term consequences if the reasoning utilized in rendering findings is faulty due to clinical motivations. This type of case usually involves conceptual innovation—the articulation of a new concept, or a new entry in a clinical diagnostic manual, or a new theory altogether. I will refer to these instances as cases of *clinical influence*.

Clinical influences are not typically outcome-oriented: the findings are not dictated. Still, these influences carry more profound risks for science. Should blowback from these types of cases be prevented? Are there any resources we can offer here for doing so? Or perhaps there are ways of managing the effects more transparently. Here are some examples to think about.

3. Flocking Boids: An Alternative Arc of Scientific Development

Consider once more the flocking of birds that we first discussed in chapter 3. This phenomenon is worthy of science. It's only recently that dynamicists have finally achieved some measure of an understanding of it. Only recently have they developed the idea—largely through mathematical and computer modeling—that the remarkable synchrony of motion among flocking organisms is orchestrated by simple adjustments of an individual trajectory, made locally but distributed widely. These can be understood as "all-over" features of a flock. Dynamicists now refer to these as "rules of flocking." Here once again is Craig Reynolds's (1987) formulation of these "rules of engagement":

1. Separation: steer to avoid crowding local flockmates in your "near neighborhood."
2. Alignment: steer toward the average heading of local flockmates.
3. Cohesion: steer to move toward the average position of local flockmates.

If the theory of local adjustments is roughly correct, flocking requires uniformly of each flock member only that it reacts to flockmates within a certain small neighborhood around itself, characterized by a *distance* and an *angle*, as measured from the organism's own vector of motion. A flockmate outside this local neighborhood can be ignored. Similar models have been devised to model the collective foraging behaviors of social insects, for example ants in an ant colony (M. Mitchell 2009). Increasingly more nuanced aspects of the "rules of engagement" are coming to light, aspects that were not present in the original abstract models—for example certain social facts (dominance hierarchies) exert pressure on the way that different organisms in the flock follow or modulate the rules (Nagy et al. 2010).

As we remarked in chapter 3, the specifics of the large-scale features of the synchronized motion matter a great deal—because flocking needs to be very fine-tuned if it is to serve the interests of the flock. How does fine tuning work? It is a matter of adjusting the specifics of the flock's "rules" without relying on any feature of any given organism in the flock. The "rules" concern "all-over" characteristics that might well be shared with flocks of very different species and may in no way depend upon the biology of the species. They are purely relations between the organisms; they are in no way tied to the characteristics of the flock members as biological organisms. Furthermore, the rules are the flock's way of reducing their degrees of freedom when in flight, and thereby they (collectively) modulate flock characteristics in a coordinated way. They are, in other words, constitutive of flock-level characteristics.

Let us illustrate again this reduction in degrees of freedom in the context of our present flocking example. Close behavioral coupling among near neighbors in a flock allows a localized change in direction to be amplified and propagated across the flock. This allows each flock member to influence and be influenced by flockmates much farther away than its local neighborhood—it gives each a much larger reach, a much more appreciable "effective perceptual range" than any one organism's actual sensory range. It is hard for groups to maintain cohesion if the coupling distance is too short, but it if too long "misinformation" can be more easily propagated, so the flock is more susceptible to error. In contexts in which high sensitivity to motions of distant flockmates is unnecessary, because immediate responsiveness to change in an environment variable is undesirable, its implementation becomes inadvisable, given that it might be costly or might expose the flock to unnecessary fluctuations.

At the risk of over-repetition, we will emphasize the point we made in chapter 3: a flock consisting of hundreds of organisms operating under a set of "rules of engagement" is decidedly *not* a System with degrees of freedom on the order of the number of flock members: it is instead a System with something on the order of a dozen degrees, counting among them rough size, coupling distance (size of the neighborhood of sensitivity), and degree of context-sensitivity, as well as environmental variables that tend to couple with these features. A flock is an entity with a reduced number of degrees of freedom than there would be without the rules. Some of these degrees lie in the environment itself, *and so cannot really be manipulated for engineering purposes*. So it is important, for the purposes of the engineering/control questions, to determine which degrees of freedom are under the control of the flock (or someone who would control it) and which are not.

What is the point of repeating this example here? Recall, we are concerned with whether there are "moves" in the analysis we have repeated, sketching once more the general shape of flocking theory, to which we must be alerted for the purposes of assessing the "bending" of Science. Are there characteristic opportunities in this analysis for misappropriations of Science, or simply for missteps? Can these be prevented, and do we have tools (preferably in the toolkit that will come with a scale-free theory of science) for implementing solutions? To answer these questions, we first have to assess how the analysis of flocking can be useful in different scientific contexts, real and imagined.

The study of flocking can teach us many things—such things as how to construct a self-transporting "fleet" of objects, for instance. We might not have known of this potential application when we began studying flocking (in actual history). But suppose that we could have foreseen this potential application from the start. Suppose that we entered into the topic of flocking with a concern for learning about controlling some kind of fleet—in the air, on water, or even on land. Would we come away with different messages or different research trajectories for a science that (however glancingly) countenances the flocking of living species as falling within its areas of concern? For instance, would an investigation concerned with (and going under the label of) *fleet control* urge scientists to stop calling groupings of birds and other species "flocks"? If so, what damage could this do?

It makes sense, of course, for "fleet" researchers to focus attention upon the facts to which a flocking analysis also responds—facts about an interlocking network of interdependent features and characteristics of "fleets": entity numbers, neighborhood and perceptual ranges, amplification, gain, information transmission times, and so forth. Some features belong to the entities (whether they are fleet or flock members) as individuals, others more properly to the aggregate "fleet" or "flock." So long as "fleet" researchers share their insights with "flocking" researchers, I cannot see the harm done. It seems that both areas of research come away enlightened, and in a harmonious way. Consider how easy it would be for "flocking" research to get taken up by the hypothetical "fleet" research we are now imagining. But the same might be true in reverse: it might have been the case that "fleet" research "spawned" an area we might only later come to refer to as "flocking research". Indeed this is precisely the kind of practice I have been preaching (chapter 2), and for which we can give credit to the unity of science.

There is considerable overlap between the real and hypothetical case we just imagined: even those concerned with the "flock" are nowise concerned with "birds"—they are instead concerned with "boids," just as those concerned with the "fleet" might not be interested in the "ship" (or what have you) so much as with its role as fleet member. True, a new term has emerged in the actual case, literally coined, and the coining is clearly motivated by important theoretical considerations. But we can imagine that this consideration would well have been shared with a discipline concerned with clinical application to fleets. This is a crucial observation. And it leads to the following profound questions: are "boids" a natural kind? Are they any different from "fleet member"? Are the former related to birds in the same way that the latter are related to ships? The truth is really that the study of flocking illuminates a new *natural kind*, and one that might easily have no obvious barriers to its applications.

The term "flocking" is now denoting an area of inquiry, and the natural corresponding entity is of course the flock. But a new independent area of inquiry (relating to "boids") has shown itself, and it seems to require new terminology for entities that are already recognized elsewhere—the members of the flock, severally known as "birds." The point of "boids," however, is to signal that flocking applies to much more than birds—and indeed that the biological specifics of birds are simply not to the point. It is "boids" that participate in flocking, not birds. Flocking is not about biological entities at all, that's the point. As it happens, "boids" have captured the imaginations of computer programmers and other artificial intelligence enthusiasts. It has thus gained currency in larger areas of the academy, and all because the nodes in the flock are now being referred to as "boids." The coining of terms is not simply empty rhetoric. It is a move in the science-making game. And science-making is indeed our topic.

Things might have gone differently. Reynolds might have conserved the resources of the mint and utilized throughout a more neutral—non-new-entity-suggestive term—like "flock member." And others might have distinguished between the different forms of flocking among birds, fishes, and herding creatures. Herding and flocking might have seemed like sister phenomena ("birds v. herds"), with mere family resemblances, not species of a single genus epitomized by the "boid." It would have been a theory about biological entities then. Or, at any rate, so someone could easily have supposed. Would this have been a better outcome? I think not. But if we assume that Reynolds was concerned with the "fleet" and not the "flock," should we find the move to "boids" problematic? Again, I think not.

The term "boid" makes it possible to see the phenomenon of flocking as independent of biology proper, as involving traits/characteristics that transcend what is distinctive of biology, but obviously these are in some sense supported by biological features. The arc of flocking theory signals a new pathway for conceptualizing a network of dependencies. Flocking theory is therefore a bridge, not a self-contained area of proprietary inquiry. In the new terms we have been employing here; flocking analysis suggests new (or at any rate different) degrees of freedom, in descriptions of flocks, ones that serve the biological entities' navigation goals and needs. This is another crucial observation. It helps us explain what bridge-building is. We should not find it problematic if we are concerned with blowback: bridge-building is the identification of degrees of freedom in ways that might be useful elsewhere. But its usefulness elsewhere cannot have a negative impact on our identifying it in the present location. Let's go back to the example to see why not.

Suppose that Reynolds's intentions are different from what they actually were—suppose he harbored certain clinical goals all along. Perhaps he wished to build speedy fleets. If that were so, should we be concerned that "clinical" perspectives are pushing Science down an unintended arc—that things would have gone differently and better for Science had Reynolds instead been concerned with flocking *for the sake of Science purely*? I think we should not be concerned in this sort of case—where the innovation is essentially a bridge. For the bridge would not do its work to spawn a new field of inquiry if in fact it were not the case that boid characteristics serve as independent degrees of freedom outside the context of birds. And this would not be true if "boid" were not well and truly a natural kind.

4. Autism

We turn next to game-changing moves in psychology. Trajectories of the conceptualizations of psychological conditions—and crucially, dysfunctions—are wildly unpredictable. As I intend to argue, they are more problematic than those rooted in engineering contexts. They are much less like the sorts of bridges we have been commending that help create corridors between different areas of study. I will take up the conceptualization of autism as my case in point.

Leo Kanner (1943) documented the very first cases to be labeled as "early infantile autism"—a term he innovated to denote a deficit in "affective contact," essentially a failure to bond with other humans. He described 11 children with striking behavioral similarities in each of three categories now recognized as definitive of the condition: "autistic aloneness" (social and communicative alienation), "insistence on sameness" (discomfort with change), and repetitive behaviors. He described intelligent children with prodigious memories who rarely asked questions or gave answers; who were indifferent to the social comings and goings around them, preferring objects to people; who were overflowing with stereotyped movements and behaviors, picked up from their environment like diseases; whose actions were repetitions, mimicries, and echolalias; who had no grasp of social proprieties; who had trouble using language to communicate; and who especially had trouble making sense of pronouns.

Kanner's own conception of the disorder focused upon the affective deficits. For him it was primarily a disorder of human bonding. Subsequent researchers diagnosed the convergence in Kanner's cluster of symptoms as a form of schizophrenia. But then in the late 1960s autism came to be established as a separate disorder of neural development, primarily by reference to its longevity—it is lifelong—but distinct from mental retardation and from other developmental disorders that affect social functioning. It is now categorized as a pervasive developmental disorder (PDD) and thought to result at least in part from genetic causes. There continues to this day robust discussion of whether there is a single explanation of the "developmental trajectory" that is now termed autism—either genetic or psychological—with much evidence thought to suggest that autism is instead a complex disorder whose core aspects have distinct causes that often co-occur (Happéet al. 2006, Happé and Ronald 2008).

This trajectory in the conceptualization of autism, as I've described it so far, seems already complex enough. We are accustomed to dealing with a certain range of issues involved in the trajectories of categories: how many "natural kinds" ("clusters of characteristics" or "clusters of causes") are being dealt with in the phenomena under study, and if more than one, how are they related to each other? But there is more complexity than this still. There are really two parallel trajectories, each competing for the label of "autism." What's perhaps most interesting to note is that they are not really seen as two parallel trajectories, or even competing over the label. Here is what I mean.

In the 1970s, research on congenitally blind children reported similarities between developmental trajectories in autistic children and children born blind. Fraiberg and Adelson (1977), for example, report that, like autistic children, the congenitally blind are frequently delayed in correct usage of personal pronouns; their general use of language in communication is delayed, they display echolalia, impaired symbolic play, difficulties with pretense, and reduced ability to produce and follow stories or accounts of other peoples' experiences. These are symptoms of impaired development of a social self. Peter Hobson (1993a, 1993b) indeed reports sensing a "challenge" to explain why the two phenomena (autism and congenital blindness) share these commonalities. Taking a cue from these similarities, he sought to explain the commonalities between developmental deficits in autism and congenital blindness as a common effect of different pathologies that each results in knocking out the same developmentally crucial element in a subject's experience. Specifically, he proposed that being blind from birth is an impediment to *acquiring the concept of a person*. And of course this turns out to be a hindrance to social connection, though not an insuperable obstacle to the congenitally blind: their deficits admit of being overcome.

Thus Hobson's diagnosis is that the biological deficits of Kanner's subjects make them susceptible to the same barriers to normal social developmental as the congenitally blind. An autistic individual suffers (as Kanner rightly observed) from impairment to biologically based capacities for emotional connection. As Hobson (1993b, 270–71) writes, this deficit, like the inability to see faces,

> compromises the degree to which they acquire a concept of persons and of themselves and others *as* persons who share the property of having a subjective mental life but who may hold separate and at times contrasting psychological orientations vis-à-vis the world. . . . [A]utistic children seem to understand the geometric properties of vision, and even relate their mirror reflection to their own bodies, but what they relatively (not absolutely) lack is the kind of *evaluative* "self-reflection" that has its origins in the experience and/or conception of being evaluated in the minds of others.

If Hobson's analysis is correct—and I find it very persuasive—it might also make good sense to say that the congenitally blind can transcend their childhood autism. Persons who are socially isolated throughout life as a result of deficits in bonding capacity never really overcome the impediments to acquiring the concept of person. They never overcome *their* autism. They are cursed with it, even if they manage to find work-arounds. On this line of reasoning, the label "autism" refers to an original state, time-indexed, that is the natural endowment of every child or organism until such time (maybe never in nonhumans) that they acquire the concept of a person—which happens routinely along a normal human's developmental trajectory without any special environmental inputs. It would appear (at least at the present time) that nonhumans, and especially nonmammals, are profoundly and irremediably autistic. And some humans struggle with it all their lives.

But no one today seriously proposes this usage of the term, as eligible as it might seem. Everyone, Hobson included, now reserves the term "autism" for a lifelong, pervasive developmental disorder that manifests in a range of developmental trajectories, in spite of strong suspicions that the syndrome cannot be put down to one single source. Instances of cases classified as autism have no one core of characteristics, but only share a family resemblance. (This of course presents a certain challenge to "find the core"—and now and again someone claims to have done so.) Still, the idea persists that we might be able to characterize autism's fundamental characteristics nonlatitudinally, possibly even in brain-physiological terms. In subsequent collaborations, Hobson, with Martin Bishop (2003), writes:

> There is substantial heterogeneity in the aetiology and clinical presentation of autism. So how do we account for homogeneity in the syndrome? The answer to this question will be critical for any attempt to trace the links between brain pathology and the psychological disabilities that characterize autism. One possibility is that the source of homogeneity in autism is not to be found "in the child," but rather in dysfunction of the system constituted by child-in-relation-to-other.

Why did the trajectory of the term "autism" take the course it did? Here's my proposal: there is a strong preference in the psychological and developmental sciences for terms that refer to longer-term trajectories. This orientation is predicated, not surprisingly, on the desire to be able to forecast and treat—on clinical concerns, in other words. A conceptualization of autism in terms of person-blindness, as theoretically attractive as I personally find it, does not serve the needs of clinicians anywhere near as well as one focused on longer-term disabilities. When a decision has to be made, the preference for diagnostics of pathological conditions and terms that facilitate utilization of treatment will usually prevail, however implicitly, even when the treatment itself is only a twinkle in the clinician's eye. At any rate, that seems to be the source of the preference for terms with medical uses. The preference might be wider in scope still—a general practical cast of mind in relation to our concept demarcation practices.

The innovation of a kind term like *autism* for something that is really (if my guess is correct) transient or at least potentially so has proven problematic. It prompts a search for something that (again if my guess is correct) does not exist as a long-term condition in every case. And it might (perhaps in other, unrelated cases) prevent identification of simple or easy resolutions to certain difficulties. In the case of autism, it seems actually to do both.

How we proceed with any given taxonomy obviously depends upon what we ultimately want to do with the categories, what we want them to do for us—but also where we project them to be leading us. And such projections also interact with our more general orientations toward our Science—how we think it's supposed to work. For the psychologist, a primary concern is with the practical treatment of patients. But of course physicists and engineers have their own distinctive practical orientations. The practical orientations of scientists always sit alongside their theoretical ones, edging and nudging. The precise balance of practical to theoretical concerns will be different for the different sciences. This fact is largely responsible for the special arc in the history of each distinct science: for one thing that distinguishes one science from the next is a distinctive balance between its theoretical and practical concerns. This balance will typically define the natural history of that science. The distinctive relationship between practical and theoretical concerns lies at the core of the sociology of each science. For each science, as we know, has a history and a sociology, in addition to a metaphysic or logic (a subject to which we turn in the next several chapters).

All sciences share the same logic, broadly speaking: that of being oriented toward the True, of being held responsible for the way that things are on the ground. But what exactly is this logic? How does it issue in imperatives to throw light upon the way things are—and in particular upon the relations of dependence between characteristics of things? This logic is in contrast with the logic of stipulation or creation or construction. A logic of creation might, by contrast, be such as to hold its subjects responsible also for meeting the needs of the stipulator, creator or constructor, or any of their patrons. The knowledge that is Science is, instead, supposed to be responsive to the way that things are, independent of the knowers' ways of grasping them or being concerned with them. This fact thus orients our attention to those elements of reality to which terminological moves are sensitive. How do certain terminological moves, as contrasted with non-terminological ones, operate within the logic of being held responsible for how things are? The answer must be this: some moves help us become more responsive to the way things are, whereas we might not need to be so responsive—or to be responsive to more or different things—when we are engaged in an enterprise with the logic, for instance, of creation. Lewis Carroll is to be held to one set of standards as the poet of "Jabberwocky" (poetic standards) and a different set entirely as the author of "What the Tortoise said to Achilles" (deductive logic), for instance. Moves have different effects in the context of different disciplinary goals. And terminological moves are an important species of moves.

So what the psychologist ought to be sensitive to, when in the throes of making a terminological move in relation to such concepts as *autism*, is the network of dependences between features of their human subjects. This proposition generalizes what scientists must be sensitive to—what they are and ought to be held responsible for getting right—is the network of dependences among the features of the entities they have chosen to study, and so of course anything that might bear on these. Scientists have to be open to dependences that they had not initially envisioned. That proposition granted, then, we can dictate no a priori restrictions on what these dependencies might turn out to be. Aprioristic restrictions are decisively, even palpably, antiscientific. They are counter to the spirit of open inquiry we call Science.

What, then, should *clinical* psychologists be sensitive to? What are they responsible to? One thing is certain: whatever else a clinical psychologist is responsible to, a clinical psychologist answers also to patients.

5. Some Modest Conclusions

We have asked: can researchers make genuine mistakes in their labeling choices? One answer supported by our considerations is that the only basis for claiming that a mistake has been made is if researchers fail to take notice of some important characteristics of their subjects that have a bearing on the outcomes in view. Researchers can never be wrong in the choice of label they affix to such characteristics as come into focus within the cross-hairs of their theoretical-cum-practical concerns. But to insist on this point is simply to affirm the conventionality of language. The researchers will end up talking about whatever they choose to talk about. This is by no means to imply that no missteps whatever are possible, because there are a number of different ways that a taxonomy can go wrong.

One conclusion from the case studies we have outlined, especially the case of autism, is that it simply doesn't matter what choices researchers (and here I especially have in mind clinical psychologists and other practical-minded researchers) make when deciding matters having to do with the usage of terms like "autism." Whatever choices they make, they will end up referring to something or other that is of interest, so long as they refer to something. So we should have no worries about whether there is a real phenomenon to be concerned about. It is the sort of concern that matters.

Now we wish to compare the two types of cases—cases in which the clinical goals were harmless (flocking) with cases in which they were less harmless (autism). I tried to suggest that in the first instance (harmless cases), the innovations were bridges to further inquiries. The bridges will often go both ways—since their existence depends entirely on whether the degrees of freedom in question are really there—at least under a reasonably wide range of situations. My suggestion is that this is a key difference between the two cases: the innovation vis-à-vis the concept of autism (that it is a single species of developmental disorder) does not travel well. One way to try to explain the reason for this nonportability is that the term is not a natural kind term. This concern is at the core of many researchers' complaints about it. They are worried that construing autism as a natural kind term prompts a search for a brain-physiological pathology that does not exist. For instance, instead of helping autistic individuals find work-arounds for a potentially temporary condition, they are worried about promoting a search for causes of their ailments that do not exist, all in the name of serving patients better.

The trouble, of course, is that if proponents of the innovation are wrong, they also won't be serving patients well at all. Of course this is not invariably true: sometimes innovations help patients even when they are not otherwise worthy. So the consequences of being wrong can take a toll on clinical ambitions, as well as theoretical ones. That clinical ambitions exist at all might create an added incentive to discover "big" syndromes that might all (one can fervently hope) be treatable the same way.

I consider the concept *natural kind* vague at best. So my diagnosis of the matter is a bit different. I will give the diagnosis first, then follow with a slight diversion on the topic of natural kinds. So, did autism get mislabeled? I'm inclined to think it did—that clinicians in their zeal for ultimately preventing or curing autism may have overlooked the possibility that it is not really a single pathology so much as the cluster of symptoms that can signal any number of possible barriers to attaining a certain capacity. This way of putting things allows me to say something further: if the notion of autism as a single, lifelong developmental pathology would not travel well, say to a discipline concerned with artificial intelligence, the reasons for this failure to port should be statable in terms considerably different from the terms characteristic of brain physiology. Interestingly, the concept of affective or bonding disorder is such a concept—as is the concept of person-blindness. But these do not require a diagnosis of a pervasive developmental disorder—a natural kind term. Indeed, the concept of natural kind seems to be the kind of thing that *should not* travel well—if it would be proprietary to a discipline, or discipline-founding. The notion of a natural kind term might in certain regards be inimical, therefore, to the idea of a unity in science. So a failure to identify a condition deserving a natural kind label, say from brain physiology, might signal a resistance to portability—something the psychologists should be wary of, given the scope some cognitive psychologists aim at for psychological science. Developmental disorders might actually be the sorts of things psychologists should wish diagnosed rarely—as contrasted with their clinical counterparts. Psychologists should perhaps opt always, where possible, for the portable terms—if only they were not encumbered with clinical ambitions. Clinical ambitions pull in directions distinctly different from scientific ambitions free of the clinic. Clinical ambitions seem to prefer physiological pathologies to diagnose—and treat with pharmaceutical remedies if at all possible or physical surgery otherwise. They show a preference for reduction. This, as I am arguing throughout this book, is a misguided preference.

Finally, could errors of this sort—if they are errors at all—be better detected if we subscribed to the prescriptions of causal modeling? I cannot think how. If these errors are going to occur, they are going to occur whether or not we construe science in causal terms. So conceding the language of causation to the clinic, as I am proposing we do, poses no special problems for the progress of Science.

Still, if errors are to occur, we should like at least to be able to characterize them—to say what is going wrong, if in fact a search for a brain-physiological pathology is not indicated (and in my view it is not). It is practically impossible to make a case that conceptualization of autism in terms of brain pathology is out of keeping with (purportedly) scientific standards having to do with causal reasoning. No such case will be viewed as plausible because at face value, reductive moves, in the present climate at any rate, enjoy a kind of gravity, a certain presumption in their favor; such moves, in the current climate, seem to have the force of scientific reasoning on their side. If the suggestion is to be resisted, therefore, it is much better to resist in terms congenial to the thesis of a unity of science I have been urging, in terms of standards that promote innovation of terminology that travels well beyond the known boundaries of a given discipline. Of course nowadays such standards are not themselves especially well subscribed. My hope is that one day they will be.

In broadest terms, the world is comprised of overlapping Systems. And whatever else they might be doing, whatever else they might be responding to, our sciences are obliged to respond to the facts of dependency among the quantities mixed up in these overlapping Systems. From a purely scientific perspective, there is nothing more to hold a given theory responsible for getting right. And there are emphatically no guarantees that we will always be right when we take a stab. But if we are working with a more practical orientation—within a clinical context—we are often held to other standards as well: we are held to the standards of service appropriate for the particular subjects under scrutiny. These concerns can pull in rather different directions. That's why Science should be wary of blowback from the clinic.

Could the same conclusions we drew about autism be true of, say, quantum theory? Is the most successful scientific theory of all time nothing but the result of a series of wrong turns prompted by a clinical or personal concern of some kind? Of course it could be. That is the nature of the scientific enterprise: tomorrow can find everything topsy-turvy. But the question is really this: has the arc toward quantum theory been the result of undue clinical pressures—for instance, political or social pressures to treat certain intellectual discomforts (patients, if you will) with certain concepts in certain ways? Could the model of inquiry itself, on which I have staked my bid for a new conception of the unity of science, be itself the result of compromised concerns? I shall have to leave it to others to render an answer, as any answer I can venture here will no doubt appear self-serving.

In the end, I think that we have to bundle clinical pressures together with "ideology." And we have to say that there is nothing to be done about either one: both are unavoidable facts of intellectual life. Fortunately, Science—conceived broadly enough to include mathematics and metaphysics—is well and truly an empirical business. Its false ideologies have to stand at the bar of experience and defend themselves. And that's how they'll be found out. This is how science differs from poetry. It's different too in that it involves a logic even more profound than that which Karl Popper referred to as its logic of discovery; the more profound logic of science is the logic of System or object integration. The remainder of this book will focus on this subject.

6

Making Sense of Order

The True Cement of the Universe

I don't want to give you the impression that I'm a great musicologist, but I'm a lot better than what I was described as for a long, long time. You know, people said I only knew three chords when I knew five.

—Leonard Cohen, *interview with BBC* (1994)

1. Five Chords

The Fourteenth Dalai Lama has a distinctive voice. I'm not talking about the quality of its sound, or his articulation of words or even sentences. Rather, I'm speaking about a certain totality of his demeanor, his presence, and his integrated message. There is a quality of interconnectedness about his person that is communicated when he speaks—something that makes it relatively easy to pick up on a certain maturity about his person, a certain orderliness to him. He is so well integrated: so many of his characteristics are conditioned by a single-mindedness.

Many other people speak with distinctive voices. Perhaps the most "voiced" people are found among the writers, philosophers, and sages; many (though not all) of them strive for some form of self-consistency—a kind of organization in themselves and their self-presentation. Voice is a kind of power—a kind of order that draws others to it. Not everyone who writes or pontificates has a distinctive voice, but the successful ones typically do. Voice can result from any number of different kinds of order—from an integrated message, or a compelling persona that conditions every outpouring, or simply an attractive way of being overall. Voice exercises power over those it reaches—above and beyond the power the owner of voice gains over self. It is worth cultivating for these, among other, reasons.

Voice is a form of self-sheathing—of organizing personal resources (and so resheathing them) in a way that allows them to function together in a more integrated way; the resources are reordered into a single, larger (and routinely quite distinctive) functionality. To exercise voice is to exercise a genuine power. It is a manifestation of an achievement: the attainment of a certain order in the universe. In that regard, exercising voice is no different from playing a musical instrument well. It is an open question—and open is how we shall leave it—whether there is a difference in kind between being able to play five chords and being able to play only three.

The conception of resheathing has manifestations at every scale of organization—as the previous paragraph illustrates. It is a conception—a principle of analysis—that transcends scale and subject matter boundaries. It is, like other principles of analysis we have examined, transportable to new areas of inquiry. Transportability, as I have been arguing throughout, is a desirable feature in a principle of analysis—it is the mark of a principle of high generality. It is a true distinction. Principles of logic and certain principles of (good) metaphysics bear this mark uncontroversially, while other principles might not carry very far. But the portability of at least some principles is one manifestation of the unity of science. This portability overcomes the modularity of the sciences—the fact that labor involved in scientific inquiry is divided between a wide range of specialisms.

When principles of order are portable, they can "seed" the field of inquiry with new disciplines. They can be implemented to create new arenas or new Systems of order at any scale. One can go from knowing just three chords to knowing five by applying either the same or further principles of order. Does this require explaining? Is this a question of scientific magnitude? I think it is. The whole System whereby layers of order fit together and interpenetrate requires explaining. This chapter takes a step—a philosophical rather than a substantive scientific step—in that direction.

2. The Miasm of the One-Scale Universe

My most general brief against reductionism—my argument's quintessential resheathing—will be presented in chapter 7. Then in chapter 8 I argue that since physics itself speaks against reductionistic metaphysics, philosophy should arrange for it to fail. Still, there is one thing I should acknowledge from the start: the world *could* have been such that reductionism is true. Nothing speaks against this being a possibility. The universe could (in the logical and metaphysical senses of "could") have been single-scale. And so nothing we will say by way of articulating a logic or metaphysic of the sciences should rule out such a possibility. In fact, the one-scale universe must serve as a limiting case (or very close to it, more precisely).

What does the limiting case look like? What would the world look like in such (as it happens, counterfactual) circumstances? And what consequences does this have for us who aspire to articulate a nonreductive doctrine? My answer is this: such a world is characterized by the absence of multiple independent schemes of order that together constitute the laws of biological, psychological, and social order. Indeed it is missing more order than that. The one-scale universe is practically order free, containing only such order as can be achieved at the "minutest" scale (wheresoever that may be judged to lie). As we move away from the one-scale universe, with successive additions of schemes of order, we find more "granulation" to the world. And so a world as rich with order as ours is much more granulated than a world with less order. The single-scale universe, by contrast, is miasm-like, a thin and nearly homogeneous mixture of whatever could lie at the "basement" of a universe (assuming basements are themselves a possibility) without scales of granulation. The single-scale universe contains just this one level of granulation.

Now, it seems entirely possible that a world could be *un*granulated—that it be *entirely* without order, not even the physical order associated with (for instance) fundamental particles. Of course this is not the world envisioned by those who acknowledge Physics as the Theory of Everything. For such a world does not contain even the bargain basement. It is clear to physicists that even this quantity of order (what we have been throughout referring to as *microphysical*) needs itself to have come about; for surely an entirely *null* universe is possible. Now I don't mean by that a universe consisting of the "quantum vacuum." I mean a universe without even a hint of order, even if it might contain entities (and they would have to be order-free). So the nonreductive proposal to which I am giving voice needs to insist upon the contingency of order itself. And, obviously, this point cannot be served by appeal to a rigid structure of echelons we depicted in the introduction and will discuss further in chapter 7—layers of order that rest on a certain privileged one. It is now clearer still how reductionism has blinded us: it has kept us from conceiving of the genuinely null possibility, seeking to work out how it might be realized, and inquiring how it might be escaped. (Astonishingly—and this is further evidence of the fact that the discipline of physics is much more advanced in its sensitivity to metaphysical matters than philosophy itself—physicists are not blind to these questions.) Just as a nonreductive doctrine should leave the question open as to whether there is a smallest scale of order, it should also leave open the question of whether there is a null universe, and what that might look like.

Here, then, is the guiding picture with which we shall proceed. We must relinquish the rigidity of a hierarchy's categories. It is surely also possible that there should not exist a biological scale of organization, and instead that the universe be organized on a physical scale, then "skipping" directly to a psychological and consequently a social scale; for these realities are surely possible in nonbiological things (like computing devices and the networks they might comprise). To make space for this possibility, one requires a sufficiently flexible framework for the organization of order itself. Sufficient flexibility in this structure requires that we forgo some form of atomism as the fundamentality dogma. It requires historicizing the metaphysical picture of order. Here is what I mean.

Fundamentalistic ontology (the atomism of chapter 3, which is present even in supervenience-based formulations of emergentism) considers the world at a single moment in time and simply assumes that its layers of order are eternal. Hence fundamental entities enter at the bottom, and the "history" of the universe is just the history of how this eternal framework gets occupied over time. By contrast, a nonsupervenientist emergentism (due to late O'Connor 2000 or Humphreys 1997a, 1997b, and some readings of S. Mitchell 2009) allows objects in the world to "run" in time and regards the order that arises in the resulting landscape as emerging from this dynamism. Both conceptions regard order as captive to the objects in which they are manifested—captive to a time capsule that spans no more than entities' lifetimes. Both these doctrines are committed to the fiat doctrine of a "bedrock" or (if more friendly toward the doctrine of supervenience of macro on micro) to a "fundamental" base of entities.

The one-scale universe is captive to the notion that the structure of world order is rigid, given, and nonnegotiable. How does order arrive? Perhaps from the beginning of the world streams of the smallest particles have had to enter this order from the bottom in order to accumulate or lodge in the preordained layers. If the structure of orders is itself eternal, we should be able to anticipate it. For instance, we should be able to say what lies "beyond" what now seems to be the highest level of order we are acquainted with thus far: the so-called level inhabited by minds or intellects organized into societies. Or perhaps you will insist that it is instead the order we know as *ecological*—of aggregates of entities mutually overlapping in a geographical region. Whichever you consider the highest level of organization, can you say what comes next? I haven't a clue. For it seems instead that order is opportunistic: it depends, quite locally, on circumstance—on what already exists and how easily new patterns might be forged next. And it can skip and branch in difference directions. The space of order is multidimensional, not single-dimensional, as the hierarchy doctrine to whose demise I am committed would dictate. Unquestionably the history of its branching is entirely opportunistic. Order builds on what already exists, but there are multiple directions in which new structures might go from any single point. If the contrary were true—if order simply increased along a single, linear dimension—then there could be no order at a certain magnitude without all the orders below also being manifest in the same spatiotemporal region. This is surely altogether too much to be knowable a priori. For instance, what barrier could exist to biological order preceding chemical order, or intertwining with it? The conception of chemical evolution—the evolution of molecules through natural selection processes—would suggest the order of these levels is indeed not rigid. And nothing we know about Physics rules out the idea that biological principles of organization precede those of Physics. Indeed, the principles of quantum theory—that most successful of all physical theories—apply to biological entities as though they were simply on a par with entities at the minutest scales. So why couldn't the latter come first?

Just imagine how things might look if order were *not* to advance along a single dimension (in the hierarchy we have been criticizing). Then how would order emerge? It would be negotiated from less-ordered things, as a new form of granularity creating itself. It would be the self-creation of order from materials already on hand, order resulting from resheathing. That seems to be how our universe has unfolded.

A universe might come into being that exhibits no biological order, no life in the familiar sense. But it might nonetheless enjoy other forms of order—chemical order, physical order, psychological order. Or there might be worlds that enjoy no physical order whatsoever, while nonetheless enjoying organic order because it simply happened to arrive first: reproduction, death, senescence, evolution, of nonphysically organized species of living entities. There is no obvious metaphysical barrier to this—what would these barriers amount to? We should not be blinded to the metaphysical possibilities simply for their distance from present intellectual economy, which happens to have been tutored in the local universe. The physical order of our own world is entirely metaphysically contingent—something that cosmologists have acknowledged for many decades—and a fact to which philosophers so often pay lip service but never really believe.

Now comes time to pay the piper. The alternative to a necessary world order is not simply no world at all (a null universe) or a world with a "different physical constants." For there might be no physics at all. Or a very different sort of physics—at least very different sorts of particles. Indeed, string theory and cosmology imagine such possibilities (and indeed some versions postulate a multiplicity of universes with different properties and sorts of order).

The ideal metaphysical framework is one that encompasses the history of the whole universe, within a more embracing context of alternative possibilities, and asks: how did order *itself* (the kind with which we are acquainted) come into being? It is thus a framework that seeks to make space for a natural history of order itself. The structure of order is itself negotiated in situ. Of course this is not the space to offer or debate answers to the question, but simply to make the question intelligible—subject to being posed—and to support science in rendering answers to it in a more comprehensive fashion. To do this, we must identify a "logic" of scientific answers to a range of questions. We do this in chapter 7.

There too we shall examine the idea that order itself has a certain *logic*—that there is a logic of order, or more properly, a *metaphysic*—and it is this: where there are laws, patterns, or order, some quantities or features of that scale are *leaders* in it, and others *followers*. And this can happen at any scale of organization whatever, from the micro scale to large-scale social order. Moreover, we must recognize the promotion of certain quantities to leadership status, and the demotion of others to follower status. We must also recognize the *reduction* of total numbers of leaders—the *reduction of degrees of freedom* (which routinely occurs in the process of resheathing). Our refusal of reduction in the present chapter will provide the proper contextualization for this *logic of leading and following* in the next. The logic of leading and following constitutes the dependence relations to which the notion of "degree of freedom" (which will be formally defined there, but which corresponds to the familiar terminology) will refer. It is the very same dependence relation referred to in science textbooks under the labels of "dependent" and "independent" variables.[1] When the logic of leading and following is complete, we will have before us a logic of the sciences.

3. The Major Transitions in Organizational Evolution: On the Natural History of Natural Laws

> So then, why is there anything except physics? . . . Well, I admit that I don't know why. I don't even know how to think about why. I expect to figure out why there is anything except physics the day before I figure out why there is anything at all, another (and presumably related) metaphysical conundrum that I find perplexing.
>
> —Jerry Fodor, *"Special Sciences: Still Autonomous after All These Years"* (1997)

How are features on different-size scales related to one another? More generally, what are the types of relations in which features (whatever their scale) might enter into? This is a very general metaphysical question about relations between features or characters—*quantities*, as scientists like to call them.

If the world is comprised of only one sort of order—one filter, such as for example in the fictional null or one-scale universes we described above—there

[1] In articulating these relations, we nowise achieve a logic of *grounding*; grounding is instead the subject of logical relationships—the relationship between a truth and its truth-makers. This class of dependence relations is well articulated by Kit Fine (1991, 1994), Gideon Rosen (2010), and Johnathan Schaffer (2003, 2009, and mss).

would be no need for cement to hold these together. There would just be one unified set (empty in the case of the null universe) of governing laws of nature. What more could one need by way of cement? Russell's (1913) notion that all we need in order to have science are laws that describe simple covariation (equations like Newton's law of gravitation, for instance) would no doubt prevail. More than that—cement, if you will—is required if there are multiple layers of order, and even more earnestly if (as we have insisted) there are multiple independent filters of different orders. But what does this cement look like?

There is no one universally accepted taxonomy of such relations. Indeed—and quite to our shame—there is no philosophical taxonomy of quantities whatever, save by reference to scientific disciplines. So, for instance, a physical property is routinely construed simply as one that enters into physical theorizing (theorizing done by physicists) in an especially proprietary way. This is of course immediately problematic, as some properties that attract the attentions of physicists—such as for example spatial separation, temperature, and many others—also attract scientific attentions very widely and feature in treatments of a very wide and diverse range of phenomena outside the range of physics. Moreover, when a property enters into a scientific treatment, there is (invariably) no unambiguous signal that the property in question is being treated in a proprietary fashion. So there can be no consensus that a common quantity belongs properly and in the first instance to one discipline employing it, but not any of the others. This being the case, it will not do simply to gesture blindly at disciplines in the insupportable hope that boundaries between disciplines can go proxy for boundaries between taxa of quantities.

The question immediately arises whether we can be in a position to postulate anything like a structure for the class of properties in advance of empirical inquiry—or at any rate, in advance of inquiry about the scientific disciplines themselves. And so what serves best, at least initially, is a taxonomy of ways in which—or a class of models of how—that class *might* be structured. The discipline of metaphysics is best served by a clear and rich taxonomy of models for possible structure of the class of properties, to guide subsequent inquiries. (The logical next step—and this is still a very long way off—would be to articulate how potential empirical findings would favor or disfavor each of the models that were put on the table.)

It is time now to give a sketch or schema of the scientific metaphysic. The sketch is as follows. Without the governance of any laws, even of Physics, there can only be a "law" of stochasticity: everything that can happen happens at the rate of chance. This is the null universe. (Remnants of stochasticity will persist even when order arises, but if stochasticity is the only "law," we are dealing with a null universe.) Was there ever a time in the history of the universe at which this law held true? I don't know. But there might have been. We can refer to this state as the *null* filter of integration—a layer one scale lower than that of Physics, and one that might have in fact preceded Physics. Now, each major transition ushers in at its point of entry into the life of the universe a new organizational scheme—a new network of properties that ultimately "hang" together, having perhaps evolved together and put "selectional" pressures on one another, as well as on properties in unaffiliated networks. The relative "tightness" and "independence" of the networks justify the autonomy of a discipline studying them. (It will not be my purpose to furnish specific stories of each of these transitions here. I simply assert that they must have occurred, and their existence is the reason we can now study the organizational or systemic features of the patterns of events that they govern. The point at this stage of analysis is simply to find the conceptual room for these independent narratives.) The independent patternings might well overlap one another in space and time (and spacetime); therefore, there can be no good reasons for prising these apart and segregating them as in a hierarchy—at least there is no basis for slotting them into noninteracting levels. Instead, we need to make sense of these overlapping, interpenetrating filters, as I have been referring to them. We require a decidedly Systems-like analysis of the matter—Systems with a capital *S*. We've already encountered the need for Systems. And there is a great deal more to say about Systems, some of which will be said in this chapter.

And so this is the metaphysical account I will be laying out in this chapter. There are many systems in the universe—systems with a small *s*. These systems are just aggregates of things—situations in which the parts stand to the whole in simple part-whole relationships. But Systems with a capital *S* are not only aggregates, but are also structured or "vertically organized"—objects with multiple scales of organization in which spatial parts stand to spatial whole in more than simple part-whole relations. Thus there are also organizational relations—relations that integrate the parts into an organized whole that holds together in ways characteristic of the laws that govern them. And organization is at the heart of a System. Thus Systems analysis lies at the heart of the metaphysical story I shall be presenting here—a story of interlocking dependence relations that make certain features of the world "hang together" in multiple discipline-characteristic ways. The analysis that will be presented is not a physical theory, but an ecumenical theory (a properly metaphysical theory because it is suitably general) of *organizational* relations between entities and their parts.[2] To attain the Systems concepts that make this analysis possible, we will require a suite of

[2] I argued in Thalos (2009) that Systems analysis is not a discipline under the broader umbrella of physics, but is something "between" physics and biology. I would now say that Systems analysis is not so much "between" as "above" the other forms of analysis. It is more properly a metaphysic.

technical conceptions. Once all this has been assembled (and the technical assembly will be undertaken in chapter 7), we will have articulated the unity of science in a nonhierarchical framework—the scale-free model of the universe.

We are accustomed to wondering about the mind and its surprising qualities. Mind (as was Descartes's conceit) is active, while body is passive. Mind is expansive— it graps, it mirrors, it becomes ever more complexly; body (at least according to Descartes) does none of that. And yet to physicists today, it is body that is the more perplexing the very concept of mass and its relationship to other physical principles lies beyond our reach. The story of this reversal in perplexity deserves philosophical attention—considerably more than the minute amount it manages to attract. The referents of "mass" and "body" haven't changed, but our orientation to what is perplexing in them has undergone a veritable revolution. Consider: body is located within a definite region of space and time. Body makes the universe granular. (Therefore its existence breaks a certain perfect symmetry.) Body has qualities of self-protection: it resists being altered. Other forms of existence—for instance waves or fields—have none of this "solidity," this concreteness, this "locality," this "gravity." Waves and fields are like wisps of smoke: insubstantial and everywhere malleable, easily added to and subtracted from in their qualities—they are linear phenomena. That makes explaining their interactions much simpler. Body, by contrast, is stolid and true, but typically not permanent. The theory of mass sits rather awkwardly within the Standard Model of particle physics—as a kind of afterthought.

So why is there something rather than nothing? Why is there matter, in addition to pure energy? Why are there objects as such—things with characteristic lifetimes, that hold together for a time before unraveling again? Add these to the more familiar questions we in Philosophy usually wonder about: Why is there a whole realm of Psychology? And why, in addition to all the objects and minds that there are, is there also sociality and society and collectivity, and all the laws governing them? Every one of these questions gestures at a major transition in the history of the universe. There are others that I did not and could not mention; I am not positioned so as to see all the transitions clearly, and it requires more than one researcher to identify them. All these questions are in the spirit of those metaphysical questions Aristotle posed and sought to answer within the scope of a general theory of Science. By embracing a unity of the sciences, in the spirit of Aristotle, we are able to recover these questions as properly scientific and to make space for those we have yet to consider.

In the same spirit, remarks on the relationship between causal dependence and other forms of dependence relations (which we will discuss further in the next chapter) fall within the space we are carving out for radical questions. So they too are remarks about the cement of the universe. And so much of what we said in this book thus far falls squarely in this area.

Not so very long ago Science was conducted exclusively by elite members of society. Scientists then, like all intellectuals and artists, were self-selected members of a class of persons systematically contemptuous of the female, the poor, the commoner, and the foreigner, none of whom were thought fit for the inner sanctum of Science. This was bad for Science because consequently scientific thought itself was systematically scornful of the poor, the low-born, the foreigner, and the female. It is no accident that the sciences that made serious advances during those eras were not the biological or human sciences; and serious advances are made within the sciences as each form of contempt is found out for what it is. This process is increasingly accelerating now that science is no longer conducted exclusively by an elite class. True, it takes an enormous social investment to diversify the knowledge corps. But we are all—and Science too—the better for it. We take fewer things for granted.

Still, it is no unalloyed good that Science is now performed by persons susceptible to public opinion, no longer contemptuous of it as once they used to be. More and more, Science earns its keep by selling wares or shares in the marketplace and defending its findings in the courtroom and the public square. As a result, it's not always good for the business of Science if its emissaries insist on opinions that rub too rough against the common grain or that counter too strongly the narratives with which the folk has grown comfortable, particularly if overthrow of folk narratives would undermine the legitimacy of sitting powers. Those in positions of power in the courts and the public square do not easily tolerate pressures to transgress their zones of comfort any more than the market enjoys forgoing profits.

And for similar reasons it might not be good for the business of Science if its representatives were to pose radical or apparently idle questions. So sometimes scientists, like other market denizens, have to compromise to stay in business: it makes the wider community more comfortable with the way they operate. Scientists thus sometimes put away certain questions for another time and another place, and sometimes even surrender them altogether, simply to maintain a fragile peace. This tightrope drama shapes Science's contemporary relations with, for example, the religious institutions of the realm. Arenas of specialism sometimes ossify into permanent boundaries of sovereignty, with the wave of a metaphysical wand dividing the world into (for instance) "natural" and "spiritual" realms, so that sensitive toes are not trod on.

Closer to the concerns of this book, Science's current institutional position in the modern social order, as a sometime ally and permanent adjunct to business and industry, oftentimes puts it at odds with a more lofty-minded academy and with academic philosophy in particular. Science and Philosophy have increasingly separated since the eighteenth century, at least partly as a result of modern institutional and financial arrangements. And damage has been done on both sides as a result. Science today is very much out of touch with its philosophical and historical roots, and philosophy for its part is divided between science boosters on one side, and detractors of science on the other. It is as though the whole business of intellectual inquiry were a football game, to be won by some, only to be lost by others. There is hardly space for a third way of doing business, hardly any room left in the academy for critical but nonpartisan regard for the condition of knowledge. This has hurt the critical intellectual process of posing those radical questions that stimulate inquiry and sustain intellectual vibrancy pervasively throughout the academy. Working hypotheses (strategies of reduction, for instance) are frozen into dogma throughout the field of scientific inquiry; and Philosophy is left to descend into inconsequential whirlpools of mere conceptual analysis because it (like other disciplines) is pressured into surrendering questions to other inquirers. Boundaries are bad for all forms of inquiry. They must be made as porous as possible in a good theory of Science.

The questions we are posing now, regarding the origins of order, are the radical questions we could not pose with our eyes upon a hierarchy of levels or disciplines—because of wariness that we might overstep a boundary. Now that we have put away the hierarchy, these questions are more intelligible, if not necessarily any better answered. And because the hierarchy is gone, we now have a better sense of what might be required to answer them. Obviously, answering these questions will help us appreciate how the world holds together. But in truth, simply asking the questions does so as well. Because asking questions, within a framework that allows them to be genuinely considered, is itself a form of progress. This is at least in part why Philosophy is so important to Science. And thus why it is so important to clear away the fog created by the false dogma of the hierarchy.

4. We Need Two Kinds of Cement

Filters provide one kind of cement to hold things together, the kind that Russell thought would displace all talk of causation: the laws of nature. Russell believed that once the language of natural laws took hold, it would displace all talk of causation. (Obviously, his confidence in this has not been vindicated.) The content of filters is discipline-specific cement. So, for instance, the filter of Biology contains the cement of order imposed by natural selection processes. And the filter of Chemistry contains the cement of order imposed by processes that combine smaller particles into molecules and allow transformations between categories of molecules. (The time scales of these processes are very different indeed, and the "laws" are also very different.)

But what about the multiple filters or clusters of interconnected laws, each largely independent of the others? What holds these together in one unified world? A theory of the cement between the filters is a theory of the world as we actually find it. A theory of *higher orders*. And a theory of science too, very broadly construed. We get two theories for the price of one. The picture of the unity of science just *is* a picture of the second kind of cement. And the metaphysics of that cement is simply the story of resheathing that I have been unfolding in so many different ways in this book. Laws and resheathing: these are the two kinds of cement in our scale-free universe. But they merge to create a single, unified sphere of ordered behavior—our scale-free universe.

Cement holds the world together. And that happens in two ways. First, quantities (the quantities/characteristics of things) interact via laws or principles (which usually belong in a single discipline, but might be portable). And second, these interactions result in regular behaviors that allow things to hold together over larger scales and longer spans of time—on a repeated basis. The latter is a resheathing of the former. And the resheathing is recursive—indefinitely iteratable.

5. Systems

Systems analysis gives us some of the theoretical tools we require to articulate the notion of filter and control, as we saw in chapter 4. And it's a strategy of analysis that makes sense of how engineered Systems (what Wimsatt 2006, following Levins 1970, calls "composed systems") function. What's more, it requires no hierarchies. Systems analysis is a strategy of analysis that draws on interlocking control characteristics.

When one performs an analysis in terms of control (the notion we introduced in chapter 4), one calls prominent attention to certain important features of the control landscape. The notion of control we have set down refers to global ("all-over") features of a System. A control analysis thus describes a System's actual state in relation to "nearby" potential states. It functions to help us track a System's relative stability in time, rather than track the specific means, in terms of forces and other features, that may (or may not) be holding it together. For certain purposes, this is important information to track—it is the kind of information never too far from the concerns of an engineer. Keeping track of such information is very different from keeping track of its material characteristics as such. Such characteristics as bear on a System's stability in time involve keeping track of its large-scale (global) relations to other things in its context.

Remember Simon's conception of Systemhood that we discussed at some length in chapter 3. In building complex Systems, he argued, one must proceed in increasingly aggregative or cumulative stages. If this multistage process is to succeed and result in a new object with novel functionalities, the result of each stage (the aggregate so far) must possess a stable structure—so as to "hold still" while the next phase of operation is being launched. Without layers of stability, complexity is unbuildable. (He thought moreover that one could expect from this process that complex Systems must typically be [*a*] modular, [*b*] intersubstitutive in their parts, [*c*] qualitatively similar with a change to their parts or their number and [*d*] stable under reaggregations of parts.) Simon was thus reaching for the idea that a System under control has to be governed by high-order structures of stability that are relatively independent of the sorts of interactions (physiological, chemical, mechanical, or what have you) that govern their parts at different scales or layers of organization. In our treatment, a System is subject to high-order laws of dynamics and control among its constituents. A system that does not obey such laws is one that quickly falls apart. As I will soon explain, this amounts to saying that a System is one in which the aggregation of the parts has undergone a *reduction in degrees of freedom*.

This point provides an occasion for an observation regarding another potentially key concept, bearing on the origins of filters: in order for "new" layers of order—new filters—to come into being, there must exist global (or at any rate scale-wide) conditions that help certain things "hold still." A membrane will sometimes serve this purpose. Membranes can create an "inside" and an "outside" and consequently the potential for control of the inside to a greater extent than control of the outside. It would seem to be a precondition for the entry of the biological (Thalos 2009). This is simply a suggestion, but prompted by Simon's own observations regarding the engineering of complex Systems. This concern for layers of stability in complex Systems comports very well with the notion of control we were at pains in chapter 4 to articulate. Indeed, I would insist that Simon had just this notion of control in mind when he talked about construction stages.

The present strategy of analysis might be construed as akin to an idea that Robert Batterman (1998 and 2000) attributes to Aleksandr Khinchin (1894–1959), the latter of whom sought to explain the successes of classical thermodynamics via a statistical treatment of microbehaviors. Batterman argues that what makes a "higher stratum" property distinct from its lower-level basis properties is simply that it's a feature of the collective basis *minus* a host of microdetails. This "subtraction" approach to the higher level allows us to conceive of these properties as multiply realizable. This is a suggestive idea. And I will discuss it at much greater length in chapter 8, as my last parry against the one-scale model of the universe. Suffice it here to say: it is different from the analysis we are pursuing now.

To understand how independence of filters comes about, we require a language of analysis that is filter neutral. Clearly, causal language won't do, for the reasons we have already detailed in chapter 4. Fortunately, we have Systems analysis, the ecumenical language of engineers. As we've already noted (chapter 4), dynamical Systems analysis mobilizes the language and mathematical technology of differential equations and brings into play the distinctive concepts of *equilibrium* and *attractor*, as well as *gain*, *coupling*, and *neighborhood*. These concepts are not proprietary property of any particular domain of objects or regime in the world. Like the content-neutral language of logic, Systems language constitutes—really and truly—a core of an ecumenical language of science; it is thus the language of metaphysics. It is not proprietary to any single discipline. But neither is it "rough" in the way that causal language is rough. Furthermore, it is capable of doing what "purely metaphysical language" (in the derogatory sense) cannot—it is capable of providing concepts (as we've illustrated) for articulation of the natural history of patterns of order.

When does a given collection of objects under a scheme of control also qualify as a System in its own right? And how can we explain this process by which a System gets built out of components? This is the engineer's fundamental query. And whereas causal language (by design) forfeits the capacity for "translating" the languages of special sciences to one another, Systems language never does so.

6. The Way Up: On Vertical Integration

Having examined general considerations regarding the process by which Systems get built, we now step back to look at the lessons we learned, and therefrom extract a formal characterization of the key elements of organization as such. The defining characteristic of the System that emerges from such normal construction processes is simply that it is built by the process—and is consequently subject to being added onto in the same way. A System is essentially an integrated thing—a thing whose elements are integrated through organizational structures or protocols. Furthermore, adding elements to an already integrated System does not automatically dissolve the antecedent integration. This is the error committed in the "linearization" move regularly performed in path analysis (see chapter 4) and why path analysis cannot give us a true picture of the cement of the universe.

How do elemental entities and already-integrated entities that may be ripe for further integration get organized further? Even to frame the question is to decline the temptation to treat any particular sort of entity or quantity as an inalienably *independent* atom—a decision that amounts to a commitment to diversity about atoms (as well as about independence). And accordingly (to repeat a point we have already made) it comes with a rejection of the idea of treating systematic correlations among these independent Systems as a matter of incidental, "external," or "ecological" interactions between them, rather than as systemic and inalienable properties of their already-prefigured arrangements—their already-fashioned, ready-for-integration systematicities.

This suggests that entities already integrated can enter into further confederation with other entities, regardless of the latter's level of integration. And in such confederations, they can retain a certain amount of independence (at least in certain of their dealings), but they can also coalesce further at important conjunctions of circumstance, forming further molecular Systems that interact in (further) complex ways with each other, as well as with and upon the original atoms. We thus arrive at the notion of Systems within Systems, of rich complexes of intricately nested and overlapping Systems. I will refer to this as "vertical integration." But I don't mean by it that integration does not occur among entities of very different sizes and scales. What I mean by "vertical" is simply that the entities created through integration will in the general case belong to a larger scale of organization than entities thereby integrated.

This vertical integration is nowise incompatible with so-called downward causation, in which a feature of an integrate can interact directly with a feature of one of its own atoms. So, for example, take the flocking discussed earlier. One can make the observation about such a case that the coupling distance of the flock has a particularly unfavorable consequence on one of the boids at the margins of the flock. There is no barrier to this statement within the present account, however false it might be in actuality. In other words, the truth of such a statement is an empirical matter, not a logical, metaphysical, or methodological one, which is as it should be.

Systemhood is a scaling phenomenon, as we've already remarked. It is the means by which entities of larger scale come into existence. Integration is the fundamental process whereby some collection of components emerges as a single entity. Granulation of the world is the result, because an entity just is, from the Systems perspective, fundamentally a bond or granule. Thus when (for example) a large number N of particles (in the billions, say) bond together to form a single rigid body, the initial number of degrees of freedom (roughly on the order of tens of billions) is dramatically reduced to a handful.

Here is a sketch of the account that will be developed in chapter 7, wherein some technical notions will be introduced:

1. When we refer to properties within a Systems framework, we are referring to something concrete and belonging to some single System. A quantity X is always the quantity of a System. There is no such quantity as "temperature" not attached to some real-time System. When we use T (for temperature), we're using a quantity *schema*, not drawing attention to a unique quantity, and at that point we are no longer discussing quantities as such.

2. A System is a network of properties (all local) governed by a control structure (protocol).
3. When Systems overlap in space and time, they characteristically *superimpose or interpenetrate* rather than layer (as they might in the rejected hierarchy model).
   a. Superimposition is not a hierarchical notion.
   b. Entities or properties in different but superimposed "filters" can interact with each other directly and nonproblematically, so as (in one scenario) ultimately to create a new filter in which both figure.

In the next chapter, we will be introducing a conception that will flesh out the founding idea that Systems are not only aggregates but *integrates*; this is the conception of *dependence*, and of course its correlative conception of *independence*. The way forward from that point will be to give some sense of how Systems are given shape, via filters, and in some cases more than one. These ideas will themselves be throwing light on points (2) and (3) above.

7. Systemic Explanation

The theory of scientific explanation has not figured very prominently in our considerations up to this point. This topic lies just beyond the margins of our concerns. For of course how we explain within a scientific discipline must bear an appropriate relationship to the way that the world hangs together.

I have offered a theory of explanation elsewhere (Thalos 2002a), whose general features might be helpfully outlined at this point. It will illuminate how a theory of explanation requires the metaphysic of the world as we have here articulated it.

On my view, explanation comprises a genus, each type of which involves illumination of a species of dependence relation. For example, mathematical explanation involves illumination of what I refer to as *functional dependence*, which comprises logical functionals as well as other systematic covariations. But there are other species of dependence relation as well, and these can be mixed and matched in the articulation of (for example) physical theories. (This has the result that the statistical patterns associated with each will also mix and match.) Finally, I show that causal dependence is not of the same type as the dependence relations to which attention is directed in physics, so it falls in a different genus altogether.

There are still many gaps to be filled in my general account. For instance, while I hold that there must be special dependence relations that the biological sciences illuminate, I do not (yet) have a characterization that distinguishes them from the dependence relations we find in (for example) Physics. The same holds good for psychology and the social sciences. I am persuaded that, while many social scientists embrace a statistical, path-analytic framework for formulating their theories, social sciences will eventually find that this model of analysis is ultimately too coarse-grained a treatment of their subject matter. In particular, psychology has a strong need to articulate the structure of dependence in which different cognitive capacities and functions stand to each other. For example, motivation is deeply intertwined with all psychological processing. This suggests that motivation is not just one more causal factor that plays just another generic causal role in cognitive processing. Motivation is a profound prefiguring condition (in the language of chapter 4, it is a *modulator*) of almost all cognitive functioning. This fact requires faithful representation. And I'm convinced that one day Psychology—perhaps with some help from Philosophy—will be in a position to articulate that representation. When it does so we will be able to state much more precisely another dependence relation that psychologists acknowledge: language processing is of huge importance for an enormous range of other human cognitive capacities. Saying "huge" of this dependency is profoundly unsatisfying. (And it's all that a path analysis can give us.) A theory of psychological dependence will ultimately allow us to make this statement more precise.

As already mentioned, forms of dependence can be combined. Indeed, it is part of the Systems theory that I have been propounding that forms of dependence so combine, and in that way to bring into existence overlapping layers of order. To say that different forms of dependence can combine to create other forms is an instance of what I wish to call *systemic explanation*. Physical explanation is an instance of systemic explanation: it is a combination of *functional* (or covariational) dependence and *ontological* dependence (see chapter 7). Systemic explanations are proprietarily metaphysical: they belong to the theory of Systems itself. And they are high-order statements in the theory of the cement of the universe.

Conversely, it might also happen that certain combinations of dependence relations never occur or never issue in new patterns of systematic dependence. Statements to that effect would be high-order statements in the area of systemic explanation. These are open questions regarding the cement of the universe.

Some other examples of systemic explanation involve reductions in degrees of freedom. Specifically, they involve a restriction on the number of System "leaders." For instance, the quantum correlations between spins of particles in the entangled systems we illustrated in chapter 1—the phenomenon that so disturbed Einstein—can be explained this way. Remember the situation: streams of particles are arriving at two very distant points in our laboratory. They are being generated in pairs, which are all prepared the same way. We can measure spin in any direction we like at the point of capture—we can open any one of three "doors," if you will. And for any "door" we open, we get a random distribution of Up and Down. But when we look at the two populations side by side, we find that if the pair of particles happen to have been measured for spin in the same direction, they are unerringly anticorrelated. When one is Up, the other is Down. This calls for explanation. Because obviously it is no accident that their spins are anticorrelated. But what explains that?

Numerous proposals can be entertained—including the proposal that they receive opposite (but determinate) spin characteristics at the site of preparation. This proposal—the apparently most reasonable—is ruled out (for reasons we need not get into again here). Classical analysis of the case then gives out: there are no more resources for explaining this quantum phenomenon. So how does quantum theory itself explain it? On the proposal I am now developing, this is explained by the fact that there has been a *reduction in degrees of freedom* among the quantities of the combined system. Among other things, spin must be conserved; so there is only one degree of freedom to be shared between the spin on the particle arriving on the right side of the lab and that arriving on the left. That means that no more than one of the two particles may take on a spin of Up in any direction. So, when one takes on the Up spin, the other, to conserve spin, must take on Down. There ends the explanation. And whereas classical explanation of the correlation is simply nonexistent, quantum mechanics generates this one. It is a very high-order regularity regarding the degrees of freedom of the combined system. And however unsatisfactory one might find it, in view of its nonlocality, it is far better than no explanation at all.[3]

What's more, these explanations, in terms of reductions in degrees of freedom, are stark illustrations of the cement of the universe associated with interpenetrating layers of order. It is very clear that a layer of order has been imposed upon things that might already obey another set of laws, holding them together in this very high-order fashion—without changing anything else about them. The higher-scale order is not achieved by monkeying with the particles' business at their original scale; it is not dependent on changing their other local features. It is simply further order where randomness might otherwise have ruled the day. Obviously this is not ruled out by logic. Nor is it ruled out in real life. The work of cement is scale free.

[3] I examine this case of quantum explanation in much greater depth in Thalos (2002a) and set it within the context of the theory of explanation.

8. The Unity of Science: The Theory of Science, Part II

As we have already noted, the unity of science was a distinctive theme of the scientific philosophy of logical empiricism (also known as logical positivism) in the early part of the twentieth century. Logical empiricists—most notably the founding members of the Vienna Circle—adopted this idea in their manifesto of "unity of science without metaphysics" (a phrase due originally to Ernst Mach). The unity of science, according to this first understanding, is constituted by its shared methodology, especially as organized around practices vis-à-vis use of language—and, in particular, around *operationalizations* as instruments that serve as bridges between less observational and more observational concepts. These practices, according to the logical empiricists, provided a way of demarcating science from metaphysics—the latter of which they derided. Unity, for these scientific philosophers, had little or nothing to do with content, shared concepts, or subject matter. And so it was a concept of unity of science around which emergentists in our time (and nonreductive materialists generally) could come together. Popper's theory of science (see the introduction) would suit them very well. A second understanding of the unity of science, also originating around the same time, and with advocates also from within the Vienna Circle, would not. The second understanding—to which we now turn—favors reductionism.

The second conception of the unity of science also has its roots in conceptual analysis—but it gestures at a stronger sense of unity than is invoked by the first conception. The stronger model of unity was endorsed by Carnap in his *Logical Construction of the World* (1928). On this stronger conception (as the reader might recall from the introduction), the sciences are unified through the unity of their subject matters, which is in turn manifested in the relations between their concepts. Roughly, scientific terminology makes reference to the world, perhaps even to the same regions of it, but different sciences make contact with the world at different "levels." And this fact corresponds to the relations in which stand the conceptual apparatuses of different sciences: the concepts of the sciences stand in hierarchy relations, corresponding to the hierarchy of the sciences themselves.[4] And how are we to have a hierarchy if there is no most fundamental level?

Today, however, philosophical culture is less averse to metaphysics as such. Philosophers of science no longer rally around "unity of science without

[4] George Bealer (n.d.) gives a penetrating treatment of the relationship between these ideas. The tension between the reductionistic and nonreductionistic ideals among philosophers concerned with the unity of science is brought out nicely in Cat (2009).

metaphysics" as a call to action. Does this open up new options for them? Yes. One pervasive conception of the unity of science today—though hardly remarked upon in these terms—draws on the concept of causation: science is held together by its concern for causal relationships, of which statistical methods (at least according to some) are often diagnostic. This theme runs through a good deal of philosophical scholarship of the last quarter century. While I am not averse to metaphysics, I have argued in the preceding chapters that this proposal for construing the unity of science is not particularly good—that the language of causation is inapt for the task of pulling the sciences together. And besides that, the language of causation is already doing something else in service of humankind—service of a more practical, less intellectual sort. Still, one might want more in the way of stitching the sciences together than merely "roughly empiricist methodology." The metaphysical planks we have fashioned will allow us to construct such a theory of unified science.

Rather than providing reductionists a foothold for their exorcising of categories of entities or characters, what nonreductivists yearn for is a rationale for embracing a diversity of categories (in advance of knowledge of how they are differentiated—indeed without regard for their nature and number) within the same form of friendly and vaguely science-approving embrace. It would thus be highly desirable to have an embracing "logic" of the interrelations among objects in different categories, without regard for the categories themselves. If only a special-science-neutral logic of such categories or their differentia could be had!

The early chapters of this book have been a means of working up to proffering such a logic, one that owes nothing whatever to the concept of causation (as generously as this term is used nowadays). This is not a logic in the propositional sense. It is, instead, essentially a metaphysics of how Systems hang together, which issues in a metaphysic of the world. It is not hierarchical vis-à-vis entities or the sciences devoted to them, nor does it provide an ecumenical language in which the sciences themselves traffic with one another (via, for example, causal concepts). That kind of traffic must be performed case by case, not via a content-neutral language. At the heart of this logic is a (partial) taxonomy of relations between entities and characters, and special diagnosis of the ways in which categories of quantities provide layers of order. Indeed, this logic is intended to illuminate what it means to say that there are layers of order. It is even more essential for illuminating why such layers are necessary. It is perhaps metaphysics in the sense the Vienna Circle disparaged. Fortunately, philosophy of science has grown (in its middle years) to have a healthy appreciation for the virtues of doing metaphysics. We are ready now for that logic.

PART THREE

DEPLOYMENTS

7

The Logic of Leading and Following

> The Greek spirit has informed the philosophy and practice of mathematics from the beginning.... The core idea is that mathematical truths are certain and universal precisely *because* they have nothing to do with the world.
>
> —David Foster Wallace, *Everything and More* (2003)

1. Something to Do with the World

This chapter is perhaps the most technical of the book. While we will discuss no forbidding examples from Physics (as we did in earlier chapters, and as we will do again in the next), here we do a bit of technical philosophy, introducing definitions in a precise way. We will repeat some points made earlier within a more formal framework. Doing this clarifies and precisifies the meaning of the terms we have been using—fixing them to our mast, so that the scale-free universe model can sail under a banner recognizable for long distances. This is the chapter in which I make good on the very first debt I incurred in the introduction: I will tell you in precise terms the philosophical meaning of my term *resheathing*. The procedure of articulating definitions in philosophy can help bridge philosophical concepts with those we find in mathematics and the sciences. The present definitions are meant by way of a bare-bones articulation of the theory of science we have been developing throughout the book.

At the core of this material there is a very simple idea: the debts and assets of the universe have to square. The currency in question is that of *dependency* or, if you prefer, *leadership*. There is a budget of dependents and another of independents—leaders and followers; the *logic of leading and following* keeps track of that information. When the budget of leaders is reduced, this is a reduction in *degrees of freedom*. When this occurs, we have new patterns of order, and (typically) also a *resheathing*. Resheathing, in leadership terms, is essentially the reorganization of old units (leaders and followers) under new units with new leadership—new management, as it were. These ideas are really at the heart of Systems analysis. This *metaphysics* has *something* to do with the world, but at a very high degree of generality. It is what it takes to make sense of a scale-free universe. (Nothing of the sort—nothing so deeply metaphysical—is required to make sense of a single-scale universe.) This metaphysic will also transform many of the topics discussed under the general head of metaphysics—a topic we touch on by way in concluding the book.

2. Basic Principles of Systems

A System with a capital *S*, as we have remarked a number of times already, is an entity subject to high-order laws of dynamics and control among its parts. A collection of entities that does not obey such laws is one that quickly falls apart. The key to understanding this fact is that a System, with a capital *S*, is one in which the aggregation of components has undergone a *reduction in degrees of freedom*. When this happens, there is an overall *simplification* (and decidedly *not* an elevation of complexity) in the proceedings. That's why it works so well to treat the resultant as a single, larger-scale unit. For example, one can treat a body made of steel or wood as a single body with a handful of characters, and not an aggregate of something on the order of 10^{23} bodies with at least as many characters. And roughly the same is true for a flock when it is flocking. We shall endeavor in this chapter to explain this idea. Once it is in place, the fundamental principles of Systems analysis, pertaining to the nature of vertical integration, will be illuminated. Here are the fundamental ideas, utilizing some basic concepts. We encountered some of these in chapter 6; I remind you of them here by way of providing the necessary context for the technical treatment of the fundamental concepts in the next section.

1. A System is a network of (local) properties standing under a *control structure* (protocol).
2. When Systems overlap in space and time, they characteristically *superimpose or interpenetrate*. Superimposition is the antithesis to the levels structure of the hierarchy in the one-scale universe model. In superimposed Systems, entities and characters in different but superimposed *filters* can interact with each other directly. Each of the interactants may go on to figure in larger filters, under further protocols.
3. This means that there is a tendency for *degrees of freedom* at lower scales of measurement to be sloughed off (in the language we used in chapter 3, they are *displaced*) and new degrees of freedom to emerge at higher scales. This is constitutive of *resheathing*.

3. Dependence Relations: A Partial Taxonomy

Our concern in this chapter is to articulate the "logic" (in the sense of high-level categorical structure) of vertical integration. This logic—the logic of dependence—will make the so-called downward causation we discussed in chapter 1 intelligible. This is the logic that will make explicit what I mean by "resheathing": the action of an entity or process at a larger scale that brings order (and a proportionate reduction in degrees of freedom) on entities at lower (and typically only slightly lower) scales.

To set the stage, we require an appropriately circumscribed terminology. For starters, we want to be clear about how we are referring to characters in the world. When we refer to properties or characters, within a Systems framework, we are referring to something local. In physical sciences, these are called *quantities*. We shall adopt that terminology here, as follows:

> *Definition 1*. A *quantity* X is always a concrete, spacetime localizable feature of some portion of the world, and of a specific System σ in particular. A quantity takes on *magnitudes* in time, and these may vary over time.

This is at least partly what we mean when we say that a characteristic of something is always *contextualized*. Thus, for our purposes, we shall not countenance such a quantity as "temperature" not attached to some real-time System; when we shall say "T" (for temperature), we shall be understood as using a quantity *schema*, not drawing attention to a unique quantity.

We wish to make statements about how certain quantities depend, not on another (single) quantity, but on a *class* consisting of a collection of other quantities. It will be generally taken for granted within our framework that statements of dependence propose the existence of one-on-one relations whose second term is a class comprising an unrestricted number of entities or quantities. Let us call that class on which a dependent term depends the class of *arguments*. (I will be using this term neutrally to refer to the items on the "right" side of a dependence relation: left *depends upon* right.) The arguments—indeed all quantities in general—will also be functions of time, so time will enter as an argument in practically all the dependence relations we will consider. So let us represent time throughout by the variable *t*, and its various instances by subscripted *t*'s.

A quantity is a characteristic of the universe that may vary in *magnitude* with time, taking on no more than a single magnitude at a single moment in time. A quantity is a *concretum*, metaphysically speaking, which possesses an identity through time; it is *not* an abstract object like the mathematician's variable. Unlike mathematical objects, or concepts, or even *general notions* such as for instance that of temperature, the temperature of the liquid now boiling in the pot upon my stove—the *quantity* Ttea water—is closely linked with the object (the tea water) to which it belongs. This connection makes the quantity a concretum.

4. Functional Dependence

I will proceed with the taxonomy of dependence by utilizing as a springboard the brilliant analysis of logical dependence put forward by Kurt Grelling (1988). Grelling, inspired by Husserl's *Logical Investigations,* undertook in the late 1930s to taxonomize the dependence relations treated by logicians and mathematicians. Incidentally, his work on this subject was published—a single article in 1988 that is the source of all the material on which I am drawing—no less than fifty years after first being presented. For, sadly, Grelling died in the gas chambers of Auschwitz, and the dislocations of war are at least partly to blame for the belated appreciation of his work.

Following Grelling I begin with a definition of functional dependence. Let Ψ be a class of arguments, all with t as argument, and let X be a single quantity, also with t as argument. Then $F(X, \Psi)_t$, to be read as "X functionally depends upon Ψ, whose common argument is t," is defined as follows:

Definition 2.

$$F(X, \Psi)_t =_{df} \forall t_1 \forall t_2 \{\forall f\, [f \in \Psi \rightarrow f(t_1) = f(t_2)] \rightarrow X(t_1) = X(t_2)\}.$$

According to this definition, if, for some t_1, every argument in Ψ (that is, every argument upon which *X* depends) takes on the same magnitude as for t_2, then *X* itself must take equal magnitudes for t_1 and t_2. (The arrow in this definition is the material conditional.) This definition captures, just as intended, the mathematician's definition of a function *X* over the set of arguments Ψ.[1] A functional dependence of the quantity *X* on the argument class Ψ guarantees that for each complete set of magnitudes of the argument class, there is a *unique* magnitude of *X*. However the reverse is not assured: there is no guarantee that for each magnitude of *X* there are unique magnitudes for each member of the class Ψ.

The arguments of a function are sometimes called the *independent* terms of the functional relation, with the implication that there is a *dependent* term as well as an independent term to every functional relation, and thus that we are dealing with an asymmetrical species of dependence. But asymmetrical terminology for the terms of a functional dependence relation is inappropriate, since functional dependence, as such, is neither symmetrical nor asymmetrical. In other words,

[1] Grelling called this relation "Equidep" and derived interesting theorems about it.

just as $F(X, [Y])_t \rightarrow F(Y, [X])_t$ is not a theorem, neither is $F(X, [Y])_t \rightarrow \neg F(Y, [X])_t$ a theorem. For some functional relations entered into by two quantities are functional relations on both sides, while others are functions on only one side. The two-sided species of functional dependence, which may be called *mutual* or *reciprocal functional dependence*, is symmetrical, for while $F(X, [Y])_t \rightarrow F(Y, [X])_t$ is not generally true, it will indeed hold for the two-sided species of functional relations. Function dependence relations are not uniformly asymmetrical. Is there some form of dependence relation that *is in fact* uniformly asymmetrical?

5. Covariation

The magnitudes of two quantities can vary together systematically.[2] Now it would seem that the idea of covariation is not captured by the formula "X and Y are reciprocally functionally dependent," for two quantities may be reciprocally functionally dependent even when both variables are constant in time, and even when one is monotonically increasing while the other changes wildly over time, taking on one magnitude no more than once. It might be desirable to have a means of capturing covariation so as to exclude at least these two problematic cases from qualifying. Let us make the following attempt. As before, let Ψ be a class of quantities, all with t as argument, and let X be a single quantity, also with t as argument. Then $\mathrm{V}(\mathrm{X}, \Psi)_{\mathrm{t}}$, to be read as "X covaries with each member of Ψ, whose common argument is t" shall be defined as follows:[3]

Definition 3.

$$\mathrm{V}(\mathrm{X}, \Psi)_{\mathrm{t}} =_{\mathrm{df}} \forall \mathrm{t}_1 \forall \mathrm{t}_2 \{\exists ! \mathrm{f}[\mathrm{f} \in \Psi \rightarrow \mathrm{f}(\mathrm{t}_1) \neq \mathrm{f}(\mathrm{t}_2)] \rightarrow \mathrm{X}(\mathrm{t}_1) \neq \mathrm{X}(\mathrm{t}_2)\}.$$

According to this definition, *X* covaries with each of the argument class Ψ when and only when for every pair of arguments t_1 and t_2 for which one and only one element of Ψ takes on different magnitudes, *X* takes on different magnitudes as well.

Covariation, like functional dependence, is as such neither symmetric nor asymmetric; in other words, neither statement (1) $V(X, [Y])_t \rightarrow V(Y, [X])_t$ nor (2) $V(X, [Y])_t \rightarrow \neg V(Y, [X])_t$ is a theorem. For there might be a pair of arguments t_1 and t_2 for which *X* takes the same value, but for which *Y* varies. So, once again, asymmetrical terminology for the terms of the covariation relation is inappropriate.

[2] Some covariation relations can be expressed by means of a single mathematical equation, for example $y(x) = x^2$. Others cannot and require the high-powered instruments of set theory. (An equation is typically an abbreviation for a universally quantified proposition; for example, "$y = f(x)$" is taken for granted as an abbreviation for the more wordy proposition "for all x and y such that conditions ϕ hold, $y = f(x)$.")

[3] This notion, too, is due to Grelling, who called it "Vardep."

However, the two cases discussed at the beginning of this section, which we would have liked disqualified as specimens of covarying quantities, qualify nonetheless under the definition just laid down, just as much as they qualify as specimens of functional dependence. For according to the definition of *V*, constant *X* covaries with constant *Y*, just as monotonically increasing *X* covaries with wildly varying *Y* if the latter takes on a given value at most once. In fact, there is even *reciprocal* covariation in these two problematic cases. However, the definition *V* also helps reveal why we should *not* disqualify these two apparently problematic cases as instances of covariation. Just like the definition for *F*, the definition for *V* is stated in terms of same and different magnitudes. And there will always exist *some*, though not always any *simple*, means of expressing how the difference between two quantities, which obey the definition of *V*, change in magnitude over time—perhaps through an equation, or some more complicated mechanism of mathematics or set theory for associating one object with another—which will justify applying the term *covariation*. For at the heart of the matter lies the fact that the problematic cases differ from the unproblematic ones only in degree of complexity of the covariational pattern. The difference between *simple* and *complex* covariations is not a difference of *category*, but a difference of degree. Hence we will not be able, qualitatively, to disallow the two problematic cases—that is to say, we will not be able to disqualify these via a definition employing all-or-nothing (i.e., categorial) concepts like *same* and *different*.[4]

Grelling writes that $V(X, \Psi)_t$ is in a certain sense the converse of $F(X, \Psi)_t$, since the following can be proven:

Theorem 1. $V(X, [Y])_t \equiv F(Y, [X])_t$.

Theorem 1 states that *X* covaries with *Y* whenever *Y* is functionally dependent upon *X*. They are locally equivalent statements. Theorem 1 entails:

Theorem 2. $V(X, [Y])_t \,\&\, V(Y, [X])_t \equiv F(X, [Y])_t \,\&\, F(Y, [X])_t$.

Theorem 2 states that *X* and *Y* covary reciprocally under exactly the same conditions as they are reciprocally functionally dependent. This fact—that functional dependence and covariation overlap in their reciprocal varieties—is further justification for classifying *F* and *V* as specimens of the *same* genus of dependence, which Grelling has with good reasons entitled *logical dependence*.

[4] Grelling considers the two problematic cases we have discussed and writes, "However, in my opinion, from such trivial cases, well known to logicians, no serious objection can be derived against my suggestion of describing one sort of dependence by the statement [F]" (Grelling 1988, 218). He does not, however, explain why not, whereas I have done so.

Combining *F* and *V* we may define the following:

Definition 4. $FV(X, \Psi)_t =_{df} F(X, \Psi)_t \ \& \ V(X, \Psi)_t$.

Grelling then shows that if *X* stands in the relation FV to Ψ, and one succeeds in keeping constant all the elements of Ψ but *Y*, then a strict correlation (which we are entitled to think of as a species of equivalence relation) will hold between *X* and *Y*. Thus we should view Grelling's achievement as that of producing a substantive answer to the (metaphysical) question: what does a correlation consist in? Answer: a confluence of functional dependence and covariation.

Now, if two quantities stand in the relation *FV* to one another, this might be evidence that one "reduces" to the other in some important sense of this term. But which reduces to which? After all, nothing prevents *FV* being symmetrical, since both F and V can be symmetrical. But the notion of reduction presupposes an asymmetry. To answer this question, we shall have to proceed with the enterprise of taxonomizing dependence relations beyond the point—this point—where Grelling left it off.

Responsibility for the direction of this inquiry, from this point forward, lies nowise with Grelling's own research directions, nor with any intimations in his own work.[5] Indeed, someone familiar with Grelling's corpus might be surprised that I have found it so useful as a springboard. For Grelling, an enthusiastic participant in the logical positivist movement, and like-minded with such as Hans Reichenbach, Carl Hempel, and Paul Oppenheim, would never have strayed far from the concerns characteristic of mathematicians. And from this point forward the analysis of this chapter shifts squarely to metaphysics. But Grelling's fertile ideas sprang from a concern to articulate the concepts of gestalt—to make precise the notions of *structure*, over and above simple notions of *form* or *shape*. The logical positivists' ambition, as we noted already, was to construct concepts for the larger in terms (quite literally, in logical terms) out of concepts for the smaller. And so these ideas prove useful in getting us part way to our destination. Specifically, they help us to find the boundaries of what we are able to accomplish with logical-mathematical notions alone. And we have now hit that boundary, as now I propose to explain.

6. Physical Dependence

We have defined the relations *F* and *V* for quantities, using definitions Grelling intended for variables (the mathematicians' abstract objects that are capable of taking on more than one *value*, though never more than one at once), since

[5] For an account of Grelling's contributions in context, see Simons (1988).

everything that goes for variables can go for quantities too. When *quantities* enter into any of relations *F*, *V*, or *FV*, they can conveniently be represented by *variables* that stand in the corresponding relations.

However there will be further species of dependence relation that quantities can enter into but which variables cannot. (As will become clear, this fact is partly due to the fact that variables are abstract while quantities are concrete.) These further dependence relations are not formal, mathematical, or logical relations; in other words, they are *not* relations having to do simply with how magnitudes—marks on a given scale, which can be compared only as to which is greater—vary over time. Rather, they are relations of dependence due to imperatives of Physics or Biology or Sociology, or what have you. Let's begin with that relation due to physical imperatives—let's call this *physical dependence*.

The correlative, and more basic, notion is that of physical *in*dependence: it is the notion of *physical degree of freedom*. The idea is this: some quantity *X* is a *physically independent* quantity whenever it is among those physical quantities whose magnitudes *shape* the physical state of a System to which they belong, and it is *physically dependent* when it is *given shape to* by other such quantities. For instance, the common statement in physics textbooks to the effect that kinetic energy depends on momentum should be read as stating also that momentum gives shape to kinetic energy, whereas kinetic energy does not give shape to momentum. If this is right, then shaping is an asymmetric relation and cannot coincide with either functional or covariational dependence. I will treat the concept of shaping as a primitive and suppose it to be governed by the following axiom:

> *Axiom AS*. If a quantity X shapes the physical state of a System σ or if it gives shape to a physical quantity Y of System σ, then it is false that X is given shape to by any physical quantity.

Axiom AS is clear that shaping is *irreflexive*, as it rules out that a quantity might give shape to itself. It presupposes that shaping is objective, since it treats shaping as a two-term relation. Thus shaping, on the present account, is not beholden to any human activities.

Practicing physicists use the terms "independent quantity" and "degree of freedom" as variants, and I shall be following suit. Before moving forward, however, I will lodge a complaint against physicists' usage of the concept. They formulate the core idea in epistemological terms: *X* is an independent quantity or degree of freedom, they will say, whenever *X* is among quantities whose magnitudes must be specified in order to specify completely the state of the system to which they belong.[6] Many philosophers of physics follow the articulation of the notion of a degree of freedom in epistemological terms roughly equivalent to this one. However, formulation of the conception indirectly, in the epistemological mode, will not do for purposes such as ours, because metaphysics is the subject of our inquiries, not epistemology. An indirect formulation such as this common epistemological one cannot distinguish between (for example) physical and functional dependence. So it might inadvertently lead to error or (worse) to confounding two or more dependence relations, without giving us the possibility of diagnosing our way out of the error. I therefore urge adoption of the following definition of physical independence:

> *Definition 5.* X is a *physically independent* quantity or a *physical degree of freedom* of a system $\sigma =_{\mathrm{df}} X$ is among those quantities whose magnitudes shape the physical state of σ.

Incidentally, different physical theories may assign the role of degree of freedom to different quantities—after all, there are rival theories, and of course we should accommodate this fact. So we may say the following:

> *Definition 6.* X is a physically independent quantity or degree of freedom of a system σ *according to a theory or scheme of representation* $T =_{\mathrm{df}}$ X is named or otherwise designated by T as belonging among those quantities whose magnitudes shape the physical state of σ.

There is therefore a clear sense to the question: which (if any) theory or scheme is correct in its designation of the degrees of freedom?

Physical dependence/independence as here defined is *not* a relation among quantities, but a property that belongs to quantities individually, albeit contextually—it is a property whose manifestation requires that the quantities to which it belongs be related to other quantities and whole systems in certain ways. We can utilize this

[6] In the *Encyclopaedic Dictionary of Physics* (Thewlis et al. 1961, 818)—one of the very few places in physical literature where there is any attempt to give instruction on usage of the term *independent quantity*—M. McGlashan defines *independent variable* as follows: "In a thermodynamic system at equilibrium $F = C + 2 - P$, intensive variables (such as temperature, pressure, densities of the phases) must be specified in order completely to define the state of the system, where C is the number of independent components and P is the number of phases." Here the term "independent quantity" never even appears, only the term "independent component." The term "degree of freedom," in comparison, receives the following definition: "The number of degrees of freedom of any mechanical system is the minimum number of coordinates required to specify the motion of that system" (Thewlis et al. 1961, 274).

conception to define the desired relation—I propose that we call it *physical foundation*. This will be a relation into which physical quantities may enter. I will give only a necessary condition for this relation:

> *Proposition 1. Y* is *physically founded upon* $[X_i]$ *in* σ (abbreviated "$P(Y[X_i])_\sigma$") only if both of the following two conditions are met: (1) all of the members of $[X_i]$ and Y belong to σ; and (2) each of the members of $[X_i]$ is a physical degree of freedom of σ, but Y is not.

It follows from this necessary condition that physical foundation is (as required) an asymmetrical relation. It also follows that physical foundation is irreflexive and objective (albeit contextual), as shaping is both. It is important to note, however, that the converse of this proposition is false, since—for example—when each member of $[X_1]$ is a degree of freedom but Y is not, it does not follow that Y is founded on $[X_1]$, for it might well be that Y is founded on some third quantity that gives it shape and is not given shape to by $[X_1]$ at all. Hence we have not presented sufficient conditions for physical foundation in terms of shaping.

Now perhaps it will be said that the notion of physical foundation I have just defined is nothing other than the despised notion of *supervenience*, according to which we can divide the world of quantities into two categories so that the following condition holds: once quantities in category A have been invested with magnitude (at a given time), the magnitudes of quantities in category B are uniquely established (at that time); in other words, there is no "freedom" in choosing magnitudes for quantities in category B once the magnitudes of those in category A have been chosen. Supervenience, as you'll recall from chapter 1, was to be the instrument that would help us avoid dualism and reclaim materialism, without falling victim to any of the evil reductionisms. The hope, writes Jaegwon Kim, was that supervenience "is a relation of dependence; many philosophers saw in it the promise of a new type of dependence relation that seemed just right, neither too strong nor too weak, allowing us to navigate between [the Scylla of] reductionism and [the Charybdis of] outright dualism" (Kim 1993, 147). Perhaps in the definitions of the last section the hopes of supervenientists have finally been realized. I intend now to explain why this thought is profoundly mistaken. My notion of foundation is nowise the notion of supervenience, for supervenience is a purely covariational notion—and really much less mathematically sophisticated than the two species of covariation relations Grelling taxonomized.

As we can now show, supervenience gets no further than any of the logical species of dependence relations introduced by Grelling for variables and reintroduced here also for quantities, because it does not move beyond treating dependence relations in terms of differences and samenesses of magnitude. Recall that the slogan "no difference in S-type characteristics, without a difference in B-type characteristics" is the guiding principle. Supervenience relations are in terms of material relations purely and so do not move us beyond the class of covariation relations—if they even manage to get us as far as Grelling's definition.[7] And so supervenience, as actually defined by its patrons, is no more an asymmetrical relation than functional or covariational dependence are. (This point was made some time ago by Grimes [1988], but it has not deterred too many philosophers from use of the notion as if it were an asymmetrical one.) By contrast, the notion of shaping is explicitly asymmetrical because it is explicitly a relation of a different quality entirely from those of the functional and covariational varieties.

The notion of supervenience has not lived up to any of its promises. If supporters of supervenience had in mind something more like my notion of physical foundation, they did not have resources for articulating it. They were playing with much simpler conceptual pieces—the logical resources available to mathematics only, which brought us exactly as far as the outer boundaries of Grelling's analysis. No one suspected that the conceptual foundations of Physics itself could help with articulating another sort of relation and so help with formulating materialism in a more profound way. But it can—and also it will help with formulating scale-freedom too, as we will see. This is rather ironic, for Physics was *not* to be the model for thinking about those sciences for which antireductionists hoped to make room. We shall show that in fact it is an ideal model!

In our present account physical dependence is defined in terms of shaping, and the latter is taken as primitive. Are we making progress or merely spinning wheels? Shouldn't we contrariwise be disappointed that the explication of physical dependence itself depends on something that is left unexplicated and might as well be called—to use a naughty, because polymorphous, term—causation? Three points in reply. First, the aim of the present account is less to make progress toward a unified account of the world than it is to articulate concepts as clearly as possible, which can be put to describing the world as it might be. In other words, the aim of my account is less to describe the world as it is than to fashion instruments for describing the world however it might be. Of course one has to have primitives somewhere. The unavoidable reason for having shaping come in as a primitive is that covariational ideas alone cannot suffice. And seeing how far one can go with covariational ideas provides a clear demonstration of what is missing. Second, I claim that the instruments I am aiming to fashion are precisely those on which Physics, as a discipline, has staked its claim. If I am right, the concepts I am articulating belong to Physics itself. (They

[7] This is true whether or not we decide to also handle all the merely potential magnitudes, in addition to the actual ones. I will not dwell on this point here, since supervenience theorists have not taken any interest in treating purely potential magnitudes.

generalize quite naturally, as we shall show presently.) For this reason—and because the paradigm of good metaphysics, namely quantum theory, plays also by these rules—the account I am advancing has much to recommend it. Third and finally: this analysis of dependence is quite different from any account of causal dependence one can hope to articulate; and we'll return to that issue in due time.

7. Some Important Relations between Functional Dependence and Physical Foundation

Does the proposition that a quantity Y is functionally dependent upon the class consisting of a single quantity $[X]$, in a system σ, entail the proposition that Y is physically founded upon X? No. *Proof.* Let "$P(YX)_\sigma$" stand for "quantity Y is physically founded upon quantity X in system σ" and "$F(Y, [X])_\sigma$" stand for "quantity Y, of system σ, is functionally dependent on quantity X, also of σ." Suppose for purposes of *reductio* that $\forall_{X\neq Y}\{F(Y, [X])_\sigma \rightarrow P(YX)_\sigma\}$. Then $\neg P(YX)_\sigma \rightarrow \neg F(Y, [X])_\sigma$ and $\neg P(XY)_\sigma \rightarrow \neg F(X, [Y])_\sigma$. But because P is an asymmetrical relation (see last section), $P(YX)_\sigma \rightarrow \neg P(XY)_\sigma$. So $F(Y, [X])_\sigma \rightarrow \neg P(XY)_\sigma \rightarrow \neg F(X,[Y])$. Hence $F(Y, [X])_\sigma \rightarrow \neg F(X, [Y])_\sigma$. However, this is false in many instances, in particular those in which X and Y are mutually functionally dependent. The hypothesis must therefore be denied: functional dependence does not entail physical foundation.

A more difficult, but nevertheless important, question is this: Does the proposition that Y is physically founded upon X, and not on any other quantities, imply that there is functional dependence of Y on $[X]$? I propose to answer this question also in the negative. The reason: it is logically possible that a quantity Y is physically founded upon a quantity X, and on no other physical quantity, but that Y, while restricted in what magnitudes it can take on by the magnitude of X, does not have a particular magnitude "forced" upon it. In other words, the magnitude of X may constrain the magnitude of Y to lie within some range, but not demand that Y take on a *unique* magnitude within that range: the magnitude of X shapes the magnitude of Y, but only up to a point. So on different occasions when X takes on the same magnitude, Y may take on different magnitudes. Hence there is no functional dependence of Y on X. Someone might challenge the sense of talking about such a case as one in which there is true physical foundation, and would therefore substitute the negative answer given here with a positive one. Even so I am prepared to call this a case of physical foundation since there is a perfectly good sense in which the magnitude of X may be said to make a contribution toward shaping the magnitude of Y. When it comes to shaping activities, a case in which Y is physically dependent on X, and on no other quantities, but is not functionally dependent on *X* is no different from a case in which *Y* is shaped *in part* by *X* and in part by something else, say *Z*; in that case we are prepared to say that *Y* is shaped by *X* but not entirely. And I think we should say exactly the same thing about the case in which *Y* is physically dependent on *X*, and on no other quantities, but is not functionally dependent on *X*. All of this is of course consistent with its being true that *in the preponderance of cases*—what I will refer to as *representative* cases—physical dependence is accompanied by functional dependence.

Consider now that case in which functional dependence is present between two quantities, but there is no physical dependence between them. We may ask: how do these cases differ from those in which the quantities are *both* functionally *and* physically dependent? More specifically, we wish to have an answer to the following question: Given a case in which a quantity *Y* is both physically *and* functionally dependent on a quantity *X*, is there a third relation between *X* and *Y* which is such that both (1) it is entailed by the relation of physical foundation of *Y* on *X*, and (2) it is *not* entailed by functional dependence of *Y* on *X*? If there is such a relation, then this may be what is added (at least in representative cases) to a case of functional dependence to bring it "up" to a case of physical dependence. It is this relation that is *missing* in cases where functional dependence but not physical dependence exists between two quantities. This third relation must admit of asymmetrical instances since physical dependence is asymmetrical but functional dependence is not.

8. Ontological Dependence

Consider the following relation, which has been alleged to hold between a work of art and its audience (cf. Smith 1988), as well as between God and her creatures:

> *Definition 7*. An entity *A ontologically depends upon* an object *B* wherever *A* cannot remain in existence unless *B* also exists.

I shall adapt this new idea of dependence (or, as I should perhaps say, old idea) in the service of formulating the missing relation discussed at the conclusion of the previous section.

I shall say that

> *Definition 8*. A quantity *Y ontologically depends upon* a quantity *X*—abbreviated "*O*(*YX*)"—just in case any state of affairs in which *Y* takes on some magnitude or other (but no specific one) is by necessity a state of affairs in which *X* takes on some magnitude or other (but no specific one).

So here is yet another dependence relation in which so many quantities may stand—for example the quantities we have already mentioned: kinetic energy and momentum, for nothing can be possessed of a kinetic energy without some or other of its parts being in states of motion.

The example of kinetic energy and momentum might suggest that ontological dependence between quantities is always symmetrical, for the converse is also true in that case: nothing can be in a state of motion without also possessing a kinetic energy. I reply that ontological dependence might be asymmetrical. Consider the relationship in which temperature stands to momentum. No system can exhibit a temperature unless some of its parts are in one or another state of motion. It does not follow, however, that a subatomic particle in motion exhibits a temperature. So ontological dependence between quantities may be either one-sided or reciprocal.[8]

I now claim that physical foundation—at least on those occasions in which it is manifested together with functional dependence—is one part functional dependence and one part ontological dependence. Physical dependence, under most circumstances, is a combination of functional and ontological dependence:

Conjecture.

$$\forall\Psi\{[X \in \Psi \ \& \ F(Y,X)_t \ \& \ P(YX)_\sigma] \rightarrow [P(YX)_\sigma \equiv F(Y,X)_t \ \& \ O(YX)_\sigma]\}$$

I do not know how to prove this conjecture.

9. Generalizing Physical Dependence

To ground the idea that Systems are not only aggregates but sometimes also *integrates*, we require the notion of a *filter*, to refer to interlocking quantities at a higher scale. That notion requires a *general* conception of *dependent quantity* and the correlative notion of *independent quantity*, by analogy with the conception of physical dependence. Once that is in place, we will be able to formulate the thesis that filters are nonhierarchically organized and interpenetrating. So here we go.

The desired (simultaneous) definitions are here:

Definition 9. A quantity X is *F-independent* $=_{\text{df}}$ both of the following conditions hold: (1) it figures among the quantities proprietary to the filter F (they are defined by that filter exclusively), and (2) its magnitudes *shape* the state of a System σ to which it belongs according to the F-protocol or control structure of σ;

[8] Ontological dependence might thus serve the emergentists in their efforts to articulate an account of "novelty." But this idea need not detain us, as we have already dismissed emergentism as overall unattractive.

Definition 10. A quantity X is *F-dependent* $=_{df}$ (1) it figures among the quantities proprietary to the filter F, and (2) its magnitudes are *given shape to* by some other quantity of System σ in that same filter, according to the F-protocol or control structure of σ.

These definitions are mutually defining: we need to state both of them in relation to one another. So the F-independent characters are the ones that are not dependent, and the F-dependent ones are those that have something (namely the others) that are dependent upon them. Notice that both definitions also involve the notion of *shaping*. Once again this is being used as a primitive. (Notice that physical dependence is a species of F-dependence—when "F" refers to physics.) We can now gloss this as the idea we have intuitively been calling "activity." A shaper is something active. A nonshaper (a character that is shaped rather than shaping) is passive. Here are our agents and patients. Notice that the notion of shaping is different from that of a force in the Newtonian formulation of mechanics we encountered in chapter 2, and still different from that of a cause we encountered in chapter 4.[9]

10. Degrees of Freedom

Filters are in no way organized into hierarchies. In fact, they could not be. Filters result from simplifications of entities at a lower scale, simultaneously with the introduction of a new entity or character on a higher scale. So, for instance, recall the Cooper pairing of electrons in superconduction, which we first encountered in chapter 1; Cooper pairs of two electrons—which are fermions—end up being physically equivalent to a much simpler object, a boson, with fewer characters overall than would be in a pair of electrons that are not a Cooper pair. To be sure, a filter comes with a protocol (a control structure that governs the relevant entities at the lower scale). So filters come with a quantity of organization (we might we call it *order*). But nowise is this a hierarchy.

A protocol, as you'll recall from chapters 4–6, involves at least a *partial* ranking or ordering of quantities. So it's important to distinguish such organization as are present in these protocols from the *cumulative* or *sedimentary* hierarchy spoken of in reductionism (particularly the reductionism favored by Kim, which we discussed in chapter 1). The objectionable aspect of cumulative hierarchies is their global structure—the set of restrictions absolutely protecting lower levels from

[9] The latter—that F-independence is distinct from causation—can be rigorously proven: I have proven it is distinct from physical dependence in Thalos (2002a) in the context of a discussion of scientific explanation. I will not abuse the reader further by repeating the proof again here.

infringing activity that might come from above. Protocols do no such thing. Quite the contrary: they are themselves the very formulas for how such infringements proceed. To put it another way, they constitute the ground rules under which *resheathing* occurs.

Recall the example of the neurons discussed in chapter 4. The protocol (Axiom Neuronal) asserts the existence of a filter that governs the activities of various neurons in the arm, and which results in χ_R prevailing over χ_{NET} when χ_A meets χ_B. This might be regarded as a hierarchy of sorts, but nothing in the protocol or its exercise prohibits the existence of another protocol that governs χ_A together with other quantities not listed in this set. In other words, nothing in the nature of protocols and control structures allows them to shield that which they govern from being subject to interactions with, or further organizational structuring and bonding to, quantities anywhere else in the space of quantities. Another way to put this point is to say that reduction in degrees of freedom via one protocol is not a metaphysical barrier to simultaneous (and further) reductions via another. One clear case of this sort of thing is when an object under one protocol (let's say a chemical one) that regulates the bonding of its electrons thus-and-so, enters also into another protocol that regulates its activities according to a (biological) thermal-regulatory regime, or yet another in which it enters an (economics) protocol as an entity eligible to make market exchanges. It is as though the economics protocol were reaching "through" both biological and chemical toward the physical. But this language of "through" is apt only if we have adopted the hierarchical, layer-cake model of quantities that we are rejecting—the model appropriate only to a single-scale universe, and which therefore makes the notion of downward causation unintelligible. There is no "through," when it comes to filters, because there are no true boundaries. We can formulate a precise articulation of this point. First, we generalize again from the notion of physical degree of freedom we defined earlier in the context of our discussion of physical dependence:

> *Definition 11.* X is a *degree of freedom* of a system σ relative to a filter F, or (for short) an *F-degree of freedom* $=_{df}$ X ranks among the F-independent quantities of σ.

At this point of our discussion, the concept of degree of freedom has been made capable of ranging much wider than the domain of Physics; it rises to the highest level of intellectual generality—to metaphysics. Physics, it seems, has been borrowing a concept that is most at home in metaphysics.

Now we can put the point we were trying to make earlier as follows: for all quantities X, if X is an F-degree of freedom, it can also serve as a G-degree of freedom. But it need not do so. When degrees of freedom are sloughed off, this is a sign that there has been resheathing. A degree of freedom in one filter is capable of serving as a degree of freedom in another. This point, which we affirm on the strength of empirical inquiry, is the formal correlate of the less technically phrased idea: filters interpenetrate. This point is a rejection of the hierarchy. It is the basis of scale freedom in our nonreductive framework.

11. The Logic of Leading and Following

I will summarize in this short section the nature of what we have achieved thus far. What we have articulated is a "logic" of the dependence relationship of "leading and following." It is the logic of quantities or characters. According to this logic, it lies in the nature of a quantity or character (if it is suitable for employment in scientific inquiry at all) that it either *leads*, or else that it *follows* a leader, in a filter-defining set of *dependence* relations or structures. The structure is, of course, a *filter* that defines or helps to define one more layer of granularity to our world. Structures are the realities that contribute to the set of natural laws, as well as the provinces of special sciences. This "logic" to which we are drawing attention now does not itself enter into the special "business" of a single natural science; rather, it is meta to all of them. It is thus properly a metaphysic of the sciences.

But neither is this logic the logic of causation—which is the logic used by statisticians to identify causal factors as either causes (causal leaders) or effects (causal followers). It is not the logic of forces (which obey Newtonian mechanics with whatsoever modernizations are required). Finally, the logic we have introduced is nowise a logic of grounding relations, such as is sought by philosophers of logic interested in the relations between truths, as such, and the conditions that make them true (the truthmakers).[10] It is not a metaphysic of logic-related matters. It is not a logic of so-called *grounding*. What's more, it is neither a species of such a logic nor does it share a fundamental concept with that logic.[11]

Key to the relationship between leaders and followers, in our logic of quantities, is an important feature of leaders: they are System *shapers*. Their *activity*, in their proprietary contexts, introduces layers of order, and arenas of organization, into which the followers can simply fall into place. Thus their role in the evolution of order is key. To achieve an even larger picture of how order comes into being, we must compare the nature of leaders-in-order with that of followers-in-order. We are still quite far from even a glance of this aerial picture of our universe.

[10] This grounding logic is the subject of work by logicians such as Kit Fine (see, for instance, 2010 and forthcoming).

[11] Contra suggestions to this effect in Schaffer (mss).

12. An Open Question

Defining the notion of degree of freedom also helps us frame a question, by way of signposting where philosophical action can move next. This question does not require settling here, but it's worth raising because of its importance in relation to the topic of scale freedom.

We defined the notion of physical dependence, and by way of it also the notion of a *physical degree of freedom*, by introducing Axiom AS—the asymmetrical shaping axiom. In the process we were able to illuminate certain statements in physics textbooks, for instance that kinetic energy depends upon momentum, whereas the reverse is not true: these should be read as stating that momentum gives shape to kinetic energy, whereas kinetic energy does not give shape to momentum. All of this was premised on the notion of *shaping* as primitive, as an asymmetric relation, and not coincident with any functional or covariational relations that are fully expressible in an equation. Here is that axiom once again:

> *Axiom AS.* If a quantity X shapes the physical state of a System σ or if it gives shape to a physical quantity Y of σ, then it is false that X is given shape to by any other physical quantity.

And now we can generalize it to the F-relative context:

> *Axiom AS–general.* If a quantity X shapes the F-state of a System σ or if it gives shape to an F-proprietary quantity Y of σ, then it is false that X is given shape to by any other F-proprietary quantity.

This axiom now stands to the notion of F-dependence as the original axiom stands to the notion of physical dependence. The result is that there are physical degrees of freedom, biological degrees of freedom, psychological degrees of freedom, and so on—degrees of freedom at any scale one might wish to propose, so long as there are reasons to suppose that there are leaders and followers in that regime, and so long as one can identify the protocols under which they participate. But now a certain new question will be impinging on our (philosophical) field of vision: Are there degrees of freedom, simpliciter? In other words, are F-degrees of freedom also simply degrees of freedom, without further qualifications, no matter what "F" is? A positive answer to the question would be at least partly justified if no independent quantity in one filter were ever to show up as a dependent quantity in another. For otherwise we will have contradictions in the (comprehensive) language of dependence—if we should allow an unrelativized conception of degrees of freedom. And I am indeed inclined to the view that if our theories, taken together, say different things about dependence—if a quantity X is appointed a degree of freedom in one filter, but dependent in another—this is only because taken together they are simply in error. In other words, that there is something rather unsettling about the notion that a degree of freedom is not a degree of freedom simpliciter. Still, there are counterindications to my inclination.

On the other side of the question, the idea that designation of the degrees of freedom of a system cannot be made absolutely or uniquely is stressed in standard textbooks on classical mechanics. According to the *New Dictionary of Physics*, this nonuniqueness is due to the fact that the "generalized coordinates may be chosen *in more than one way*" (my emphasis). How so? If we wish to specify displacement of the center of a billiard ball, say, from a certain reference point within a closed room, we can do it in (much) more than one way. We can specify rectangular coordinates of a vector with tail anchored to coordinates of the reference point; rectangular coordinates will be vertical distances of the center of the ball from planes that intersect at right angles at the reference point. We will, of course, require three coordinates to make the specification in a three-dimensional room. Or we can designate the center of the floor as reference point and one ray lying in the floor and emanating from that point as direction of reference, then specify two angles (an azimuthal angle and a vertical angle) and a distance from the center of the floor. And there are as many more schemes of specifying displacement as we might care to entertain. Specification of all six coordinates mentioned above will reveal that (at least) three of them will be excessive, because (at least) three will always covary with the others. If the object whose coordinates we wish to specify is constrained to remain on a particular surface, say a model train track, then we will recognize that even three coordinates is excessive, and we can make do with fewer yet (for instance, merely distance along the track from a certain point will pinpoint the caboose). The smallest number of coordinates required to specify the location of an object is that object's number of degrees of freedom. And these kinds of examples are multiplied exponentially in quantum mechanics.

So is the question of whether degrees of freedom are absolute to be settled the same way as we settle the degrees of freedom associated with spatial position in three-space? I leave it with the reader as an open question.

13. Resheathing and the Cement of the Universe: Recap, with Technical Terms

Thus far it might appear as though the principles and definitions we've selected could be empty. In other words, it might seem as though this is an empty exercise in armchair metaphysics, motivated entirely by a desire to do away with an ugly hierarchy. Do we have reasons to believe that the principles and definitions actually hold good of the world we live in? The answer is yes, a *big* yes. In the next chapter, we will show that in fact these principles underlie our best physical theories. Physics—across a very wide range of theories from those meant for the small to those meant for the very large—is a microcosm of the principles we have been laying down. Indeed, our principles constitute a unification of Physics, in one understanding of that term (though not the usual one, for instance as construed by string theorists). And so, if the principles unify Physics in our sense, it is my hope that they unify Science very generally. What these principles do, philosophically speaking, is help us to make sense of the two types of cement in the universe. First, there are the F-specific degrees of freedom. These identify the cement between leaders and followers within a given filter—the rules of the protocol articulated by their constitution as a System under that filter. Then there are the reductions in total (combined) degrees of freedom over a number of scales when resheathing occurs. These are indicators of the cement that structures Systems vertically—creating new layers of order on top of existing layers. There is no reduction *tout court*—no loss of total number of quantities; the reduction occurs in the total (combined) degrees of freedom. A theory of Science has to have room for a distinction between these two types of cement.

14. The Theory of Science, Part III

> Dogma. Everything has a physical manifestation. So everything is—in some sense of "is"—physical.

Physicalism has been with us since Physics began enjoying meteoric successes in the academy. Physicalism in its most innocent form, and in a form in which it is objectionable only to the most confirmed dualist or spiritualist, is the thesis that every empirical entity has a manifestation—or as I will refer to it, an *image*—in the physical world. Hardly anyone has quarrels with, for example, the idea that a biological organism has a physical manifestation, a physical "aspect" or "footprint." (The dualist has quarrels because the dualist insist on the independence of another substance.) What I object to is characterization of that image or footprint as the entity's "root" or "fundament," its "rock-bottom reality." So clearly my quarrel is with the inference in the two-sentence argument I indented above—what I labeled "Dogma." Clearly this argument presupposes one and only one notion of fundamentality (hence its dogmatism) and one notion of the basement of the universe, both of which I have already resisted.

There is, however, something further that this dogma is capable of doing—something that is in fact worth doing. It presents us with a conception of the unity of Science worth having: the unity of Science, on this conception, rests in its being inquiry into the "same stuff." This is a conception of the unity of Science that is much worthier than the conception of Science as unified simply through a methodology or logic of confirmation, or as the application of statistical tools for diagnosing causation. Science, on my preferred view, casts a light on the one—the only—world.

One way of uniting the various branches of science under the flag of "light-caster" is in accordance with the Master Science model. Broad, in the early twentieth century, characterized the foundation of this union as follows—a characterization I also quoted in the introduction:

> [There] is one and only one kind of material. Each particle of this obeys one elementary law of behaviour, and continues to do so no matter how complex may be the collection of particles of which it is a constituent. There is one uniform law of composition, connecting the behaviour of groups of these particles as wholes with the behaviour which each would show in isolation and with the structure of the group. All the apparently different kinds of stuff are just differently arranged groups of different numbers of the one kind of elementary particle; and all the apparently peculiar laws of behaviour are simply special cases which could be deduced in theory from the structure of the whole under consideration, the one elementary law of behaviour for isolated particles, and the one universal law of composition. On such a view the external world has the greatest amount of unity which is conceivable. There is really only one science, and the various "special sciences" are just particular cases of it. (1925, 76)

This is not merely physicalism—a preference for the substantial and observable over the private or "spiritual." It is fortified with a preference for Physics, with Physics as Master. Broad rejected it, for reasons related to mine—namely that it implies a one-scale universe, which he did not recognize as his own.

But if one is not to draw out a conception of the unity of science in terms of a master "fundamental" science that casts light upon things at a certain scale only, then how is one to draw it out? I propose we rely on the alternative notion of fundamentality we articulated in chapter 3. I propose we instead think about how dependence relations (as we've spelled out in this chapter) contribute to integrate ingredients into the objects they constitute. Rather than the unity of Science being manifest in the fact that everything that is worthy of the name throws light on "the physical" (meaning of course "the micro"), we should instead think of sciences as unified by way of all of them throwing light on the dependence relations that are manifest in the way the world in integrated. So Science explains phenomena by throwing light on how dependence relations give rise to regularities. This is a statement of high generality indeed and describes every form of scientific inquiry that today goes by the name—even mathematics.[12] Something that enjoys scientific credentials proceeds by throwing light on dependence relations—and in particular, the reduction of degrees of freedom via laws or some such, perhaps a resheathing process. For example, the theory of Cooper pairing throws light on superconduction, by explaining how electrons (which are fermions) are brought together into pairs that behave instead like photons (bosons) and therefore crowd into the (single) lowest energy level available to them. There is a dependence relation being illuminated in the reduction of degrees of freedom implicitly in the Cooper pair: two electrons (according to the theory) come to resemble something with much fewer degrees of freedom than the original pair, and whose properties are in line with the phenomenon to be explained.

Notice that the conception of unity spelled out here is a conception without hierarchy. Similarly, the theory of explanation privileges no single science in any way. This is the best form of unity of science—a unity without hierarchy. It is of course a theory that embraces each science as autonomous—since none is master. By contrast with our proposal, the unity of science as orchestrated by a Master Science is sometimes, ironically, also thought to be in service of the autonomy of each science. But we should not be deceived: any notion of a Master Science is in fact inconsistent with the true autonomy of other sciences. Now the autonomy of the sciences is, on my view, a simple truth, known by observation—but also requiring some explication because different sciences often make mention of the same characters and objects in the world. But I do not defend this simple truth by doing homage to Physics as "more fundamental." Such deference is a betrayal of autonomous sciences. The better explanation of why sometimes different sciences make mention of the same characters is that filters overlap. The better ground for defending autonomous sciences is that there is a unity to science, but a unity without masters or slaves. Just as a union between unequals is a shabby excuse for a marriage, so a union of unequal sciences is a shabby excuse for a relationship between autonomous disciplines. I propose unity of science without masters or slaves. I propose a society of equals.

15. From Atomism to Systemism: The Theory of Science, Part IV

The presentation in this section utilizes some mathematical terminology (technology from the theory of functions and category theory), as well as some drawings involving circles and arrows for ease of presentation. The pictures

[12] I have discussed this conception of explanation elsewhere (2002a) and how it does better than its rivals to capture philosophical work that we need explanation to do. I won't repeat this analysis of explanation here.

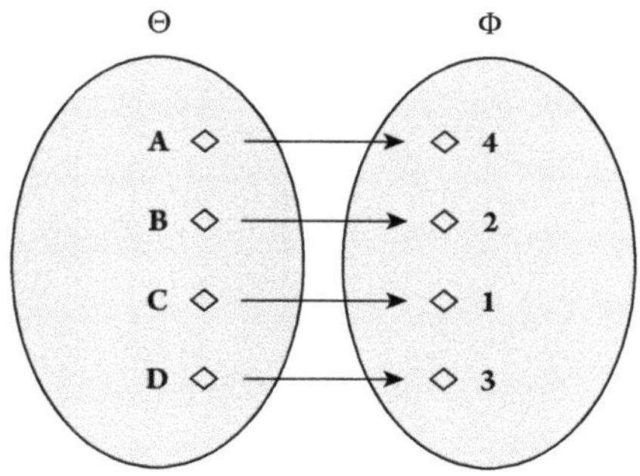

Figure 7.1 Bijective realization function *f*

help make the philosophical point I have been hammering—the "no hierarchies" point—very simple to argue for. They also help to make clear how the vision of science I fervently hope will prevail someday is quite different from that vision against which I have mounted an offensive. The conclusion of this section has already been reached in previous chapters, and readers who wish simply to get on with things can skip this section and proceed to the next chapter. Here, before we proceed, is the conclusion, stated without the props: there are no hierarchies of characteristics or entities in nature, and so no hierarchies among the sciences.

For those who wish to view the props, consider figure 7.1. It represents a *mapping* between two terrains. The arrows of that figure are all based in the space Θ (what is referred to as the *domain* of the mapping) and all terminate in the space Φ (what is referred to as its *range*). No two arrows begin at the same element of the domain, and no two arrows terminate at the same element of the range. The arrows thus represent, in mathematical terminology, an *injective* function—call it *f*. Furthermore, the arrows leave no element of either range or domain unmapped. And so the arrows also represent a *bijection*.

Suppose that we interpret the domain of this function as the psychological world, and the range as the physical world. Elements in each, let's suppose, represent properties or qualities in each world. In language we have already employed, Θ (the psychological world) possesses an *image* in Φ (the physical world)—a physical aspect or realization. And this image is made explicit by the function *f*. When such an *f* exists, it can be viewed as a *realization* function. The function in this figure is special, as we will now discuss.

The situation depicted in figure 7.1 is the best case for *reduction* of the one category to the other. It gives us the best metaphysical basis for being able to treat the terms on each side as linguistic variants, each one of the other. Remember: the point of reduction is to insist that there are not two categories, but only one, with a surplus of labels (which prompts a need for decipherment). A reduction *to the physical* in particular is a *further* insistence that the surviving category must

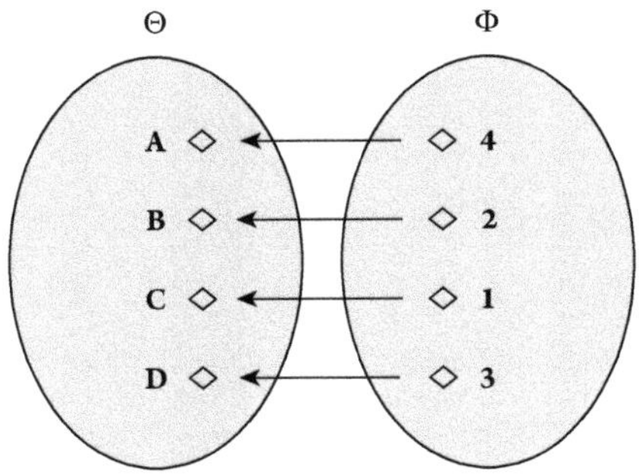

Figure 7.2 Bijective functions have bijective inverses.

be the category of the *physical*. Thus reduction to the physical requires an asymmetry between the original categories.

But now here's a simple point of mathematics: the categories in our figure 7.1 are not asymmetrical: there is no special priority attaching to the range where the image lies. True: the *arrows* imply a directionality. But f is a bijection if and only if its inverse (f^{-1}) exists and is a bijection too, as in figure 7.2. Merely reversing the arrows represents that function. And this inverse (f^{-1}) has all the credentials that f itself enjoys from the point of view of reduction. And so the case for *reduction*, under the very best of circumstances, is incomplete. It is no true reduction which might just as well have been run in reverse. (This argument bears some resemblance to that on offer in Hüttemann [2004, 7] against what he calls "microphysicalism," the "ontological priority of the micro-level." But Hüttemann has no objections to the larger thesis that the physical at all size scales enjoys ontological priority—the thesis I'm complaining so vociferously about.) It is at its very best a case for *simplification*—a pruning of an superabundance of labels—and not a decisive case for eliminating a certain set as inapt. It is at its very best a case of too many shoes; it is decidedly not a case for which shoes fit better.

Even so, one might suppose that a case can still be built for asymmetry. The asymmetry might result, not from properties of the functions themselves that connect any two given categories, but rather from the aggregate or global pattern of such functions among the pairs of the entire class of categories. For instance, suppose it is true that for every category α in the class of categories $\aleph$—let's just use as an example Ψ, the biological world—there is a bijective function from it onto Φ, but that this feature is not true of any other category. So, for instance, there might be no bijection from Ψ to Θ. The picture is what's shown in figure 7.3.

Physics "penetrates" all the other sciences. This is not true of any other science; hence Physics must enjoy a special "fundamentality"—and this fundamentality is the basis for its status as Master Science. But there is a problem with this argument. Bijections compose: so if f is a bijection from Θ to Φ and g is another bijection

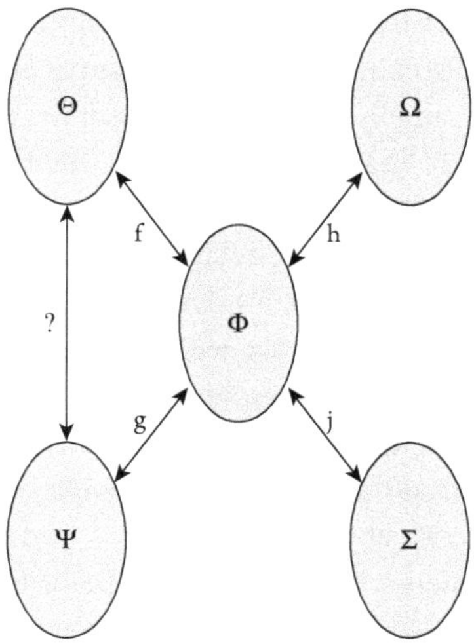

Figure 7.3 A would-be structural asymmetry

from Ψ to Φ, then $k = g^{-1} \circ f$ is a bijection from Θ to Ψ. The asymmetry drops away: we are now in a position to redescribe equally well the situation as one in which Φ is no longer the "fundament" and something else is the penetrater of all. Symmetry is restored. There is no longer anything special about Φ. Since it is easy to find arrows that connect the exterior four ovals in figure 7.3, there is nothing structurally special about Φ in this network. Fundamentality—if it exists in the form that those who talk so much about it wish for—is nowhere near as apparent as the intuitions that cling to it might wish for.

So far in these exercises we have granted (simply for purposes of argument) the existence of bijections from at least some categories α in the class of categories ℵ, to Φ (the special category of Physics). The suggestion that such bijections exist might be extracted from our drawings. But this suggestion is itself hugely problematic. In fact, bijections between categories are rare to nonexistent. Arguments to the effect that mappings such as are suggested in figure 7.1 exist are thus very problematic. Indeed the emptiness of the set of categories with bijections into Physics is an oft-discussed—and so obviously nontrivial—thesis. The emptiness thesis is the basis for the many suggestions to date that no special science is straightforwardly reducible to Physics, contrary to simple postulations (such as those of Carnap, whose views we discussed early on in this book).

There are nowadays few defenders of figure 7.3 as the model of the unity of science. Philosophers have generally come to accept the notion that monism may not really be eviscerated by failures of reduction. Indeed, many philosophers now contend that failures of reduction might be explained, entirely in physicalistic terms, if the footprints looked more like those in figure 7.4 than those of figure 7.2.

In figure 7.4, we can see that the realization map from Θ to Φ is not a bijection: it fails to preserve distinctness (mapping distinct elements of its domain to the same element of the range), while leaving elements unmapped at both ends. (This picture is essentially the idea of so-called "multiple realization" that we encountered in the introduction.) These "complications" play havoc with the possibility of reduction, but as recent metaphysical arguments have it, they need have no impact on the aspirations for a spare ontology.

Nowadays friends of physicalism are replacing the thesis of reducibility with a simple and more straightforward ontology, indeed simply with the picture in figure 7.5, which speaks for itself via the labels on the right:

This picture simply makes a statement to the effect that the only independently existing entities exist at the "ground floor," and everything else is composed of these as constituents, in ever more increasing levels of organization. As long as we stipulate, as in this picture, that the ontologies of the "nonfundamental" levels be derivative, we can fulfill the promise of monism. Physicalists whose

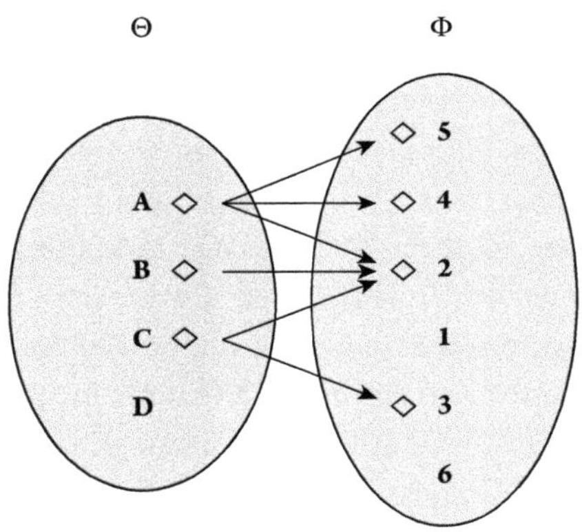

Figure 7.4 The true unity of science

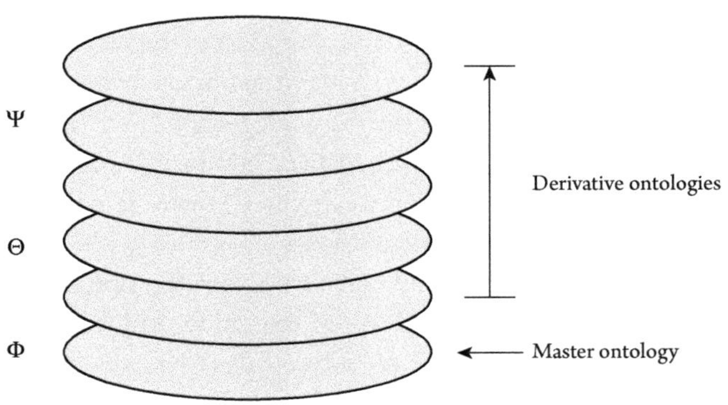

Figure 7.5 The contemporary reductionist ontology

primary concern is for preserving monism do it, via this picture, simply by insisting (as an additional, independent thesis) on the priority of the ontology of Physics.

To make good on the promissory note that everything falling in the "nonfundamental" category is derivative, one has to go further than simply providing a picture. One actually has to develop a philosophical notion of "derivative"—to articulate precisely the sense in which what is nonfundamental is different from the fundamental, but still in some sense legitimate as a target of scientific scrutiny. For some decades now, proponents of so-called nonreductive physicalism have sought to do this. Enormous intellectual energies have been poured into notions like that of "supervenience" and "emergence" in service of this goal. Metaphysicians in the supervientist and emergentist camps both share the picture of figure 7.5. Indeed each camp sees itself as this picture's best champion.

Chapters 1–3 of this book have laid out the case for the ineffectuality of the arguments for the priority of a "basement" level. But let's, for the purposes of what remains of this chapter, grant the distinction. Where does supposing that we can make a principled distinction between the derivative and the nonderivative lead? We can take measure of this very swiftly.

Consider this: whatever "derivative" amounts to, it will apply to many realizational structures that resemble those in figure 7.4, such as for example the one in figure 7.6.

Figure 7.6 depicts some multiple realizational relations, which are just as evident among physical characteristics as anywhere else. For example, a quantity of energy may be realized in many ways, and indeed in an indefinite number of ways. The bulk properties of matter generally enjoy a multiplicity of ways of being realized—as we illustrated in the example of the desks that figured in the catastrophic flood narrative in the introduction. Just for a simple example, take

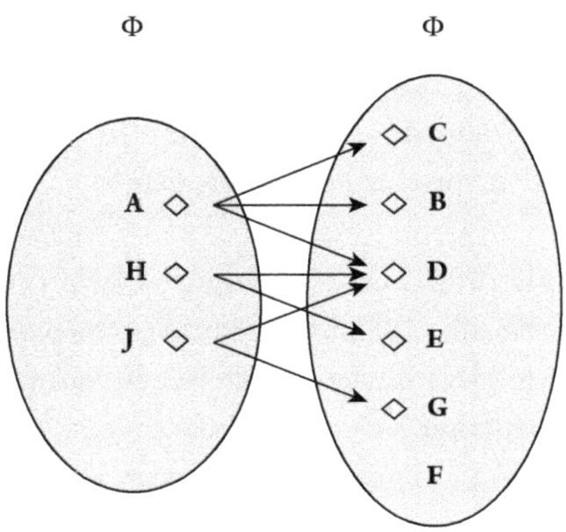

Figure 7.6 Multiple realizational relations

the property of being massive to the extent of two kilograms; this property can be realized in wood or in protein, for instance. (I will talk at much greater length about relations among physical properties in chapter 8.) The logic of the argumentation in favor of the model of figure 7.5 would then dictate that we treat the characteristics on the left-hand side of figure 7.6 as ontologically derivative.

But if some physical characteristics are derivative, then being a character featuring in the theories of physicists can no longer be construed as the distinguishing mark of nonderivativeness. This opens up a fair question: why couldn't some nonphysical characteristics also qualify as nonderivative? After all, if belonging to the discipline of Physics is not the criterion of nonderivativeness, then why couldn't a characteristic that is manifestly biological be nonderivative? For we have just established that, because of the fact of multiple realizability everywhere, we can no longer answer such questions by consulting the map of disciplines as depicted in figure 7.5. So if the program of articulating a philosophical conception of derivativeness succeeds, it spoils any argument for the map of figure 7.5. (The observant reader will recognize this argument as a generalization of the argument in the introduction.)

It is indeed ironic that the fact of multiple realizability has undermined this form of nonreductive physicalism. For it is in its broadest features the argument that nonreductive physicalists (the camp in which dwell fans of monism that oppose reductionism) typically use against their reductive opponents. This fact is itself quite revealing—it is just one more illustration of how similarly unstable reductionism and nonreductive physicalism are. As John Bickle writes in the *Stanford Encyclopedia of Philosophy*:

> At present, nonreductive physicalism is (probably still) the dominant position in Anglo-American philosophy of mind. Its proponents appeal to the multiple realizability argument to challenge all versions of psychophysical reductionism and identity theory. Ernest LePore and Barry Loewer (1989) have even called the nonidentity of mental content with any physical properties "practically received wisdom" among philosophers of mind. They cite multiple realizability as one of the principal arguments for this conclusion. (Bickle 2011)

Long story short: the picture is inapt. We require a better one. I propose that the picture early emergentists, like Mill, had in mind of the world is more like that in figure 7.7 than figure 7.5. (This image is scale free in more than one way, but that point will not detain us further.)

In figure 7.7, the size of an image of one category in another is represented by the size of overlap—the topic areas or concepts that are shared between them. So Physics might be distinguished by the size of its overlap with other categories.

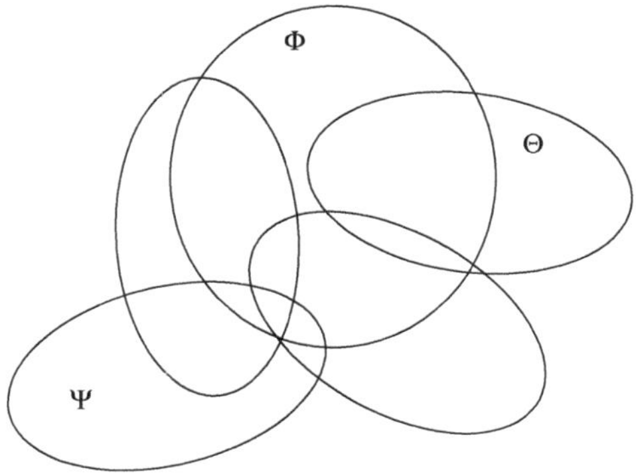

Figure 7.7 My picture, no hierarchies

But overlap, in this picture, has nothing whatever to do with derivativeness. Elements outside the overlap areas represent the "new" elements that the discipline proposes to discuss—the new topics or concepts that are not shared with other disciplines.

The world we live in, or so I maintain, is the world of figure 7.7. It is a world in which new things and new types of order can come on the scene at any time. It is a democratic order, not a hierarchical one.

8

The "Microcosm" of Unity (of Science) That Is Physics

> One is driven into reductionism when one is not cultivated to possess an array of distinctions rich enough to let things be what they are. In contrast, making the decisive distinction has an illuminating and liberating effect because it lets the concrete occurrence stand forth for what it is. We understand it not in terms of a decipherment, but on its own terms.
>
> —Robert Sokolowski, *"Making Distinctions"* (1992)

It should be no surprise that Physics is a microcosm of the sciences, seeing as it embraces a vast range (quite possibly the full range) of scales. At very small scales of measurement we discovered that physical phenomena violate supervenience, because locality is violated all the way down to the lowest scales. Nobody proposed we stop calling what happens "down there" physics. Instead, it became appropriate to employ new models—synced or linked—of physical phenomena. We changed, accordingly, the way we conducted the business of the discipline—changed our attitudes, changed our conceptions—embracing in the process, however implicitly, the reality of a scale-free universe, because the discipline of Physics manifests enormous unity, a unity that is meant to do justice to its subject matter, the universe, and because the world itself is a single unified entity. Everything that exists is physical, but then, so is everything else! (The statement is not very informative. It only seems informative because there are those who apparently do not accept it, but so long as we're not talking about ghosts and ectoplasm that no one else is meant to see, there's no reason for them to be disagreeing with us.)

Physics (as a discipline) exhibits unity in spite of all the different layers of order we have discussed in the previous chapters, and all the different models embraced in service of making sense of it all. It harbors multiple interpenetrating filters. And so the unity of Physics is the unity displayed in my model of the scale-free universe. What makes Physics so brilliant, from a philosophical point of view, is that it is able to discuss its own layers of order, and in doing so to speak of the scale-freedom of the universe using the deep grammar common to all of science. Physics is in that way profoundly philosophical.

Physics is a very lively intellectual discipline, but its inquiries are not often conducted in terms of concepts accessible in a natural language. (It serves to bear in mind too that the mathematics of one physicist is not always accessible to another physicist, any more than it is accessible to a mathematician or computer scientist doing something completely different.) This should not be a barrier to communicating the lively ideas. Because ordinary language (like language in every specialized context) can be enriched accordingly, so that it might "reach up" to and make contact with ideas propounded in Physics. Bridges must be built if the uninitiated are to become initiated. Models or analogies can help, for instance the elegant models Steven Strogatz is so brilliant at articulating. (Remember watermelon physics?) But language to articulate the key elements of the models is required—and should be made as precise as possible. In my estimation it is among philosophy's tasks to build the bridges between science and ordinary language—to innovate language that captures the key elements of a physical model or the "moves" made in a physical argument (for instance the moves in Schrödinger's derivation of his equation). I earnestly hope that this book builds some useful bridges.

Physics is also very self-reflective. The brilliant mathematician Kurt Gödel made logic and arithmetic speak about themselves, so as to say whether certain of their own statements were provable in-house. Similarly, physicists are making Physics speak about its own models, to articulate the possibility of their scale freedom. And that's what I wish to discuss in this chapter.

1. The Vast Ocean of Scales That Is Physics

Physics deals with objects at the very smallest scales we have managed to conceive, as well as objects at the very largest. If any discipline has had to deal with how to construe the relationship between scales of order, Physics has. Studying the entire range of subdisciplines of Physics is thus itself an exercise that is enormously illuminating.

In previous articles I have discussed a broad range of treatments in Physics in relation to this topic (2002 a and 2006). I have found, particularly in specialties that deal with the bulk properties of matter, that there are theories that strongly resist characterization as treatments of macrobehavior exclusively in terms of behavior of microconstituents. For these are treatments in which macroquantities are cast in the role of genuine and irreducible degrees of freedom in the sense articulated in the last chapter. I referred to these theories as "nonreductive physics." I said in these articles that Physics is in general nonreductive. But this is a negative characterization. If Physics is not reductive, then what *is* it? This book provides the answer: Physics has become reconciled to the scale-free universe. I want in this chapter to further emphasize this point.

We have already arrayed a number of theoretical cases that illustrate or are emblematic of the scale-free universe model. Many of these cases invoke principles from quantum theory to explain a phenomenon, some of which were perhaps unfamiliar to the reader. Most involve sync. We started in this book with sync as the problem (remember the lab setup in chapter 1) since, at least in the first case we analyzed, it demands forfeit of that most sacred of cows—locality. But then we turned things around, and sync became a *solution*—we saw how it helped with making sense of superconductivity, through the process of Cooper pairing. It also makes sense of lasers. So sync modeling is our friend. And since it involves loss of locality, it suggests that there is a scale freedom to the universe. We saw that scale freedom could be construed in a way that is not so unfriendly—locality could just be construed relatively: what is construed as remote from here at one scale of measurement, can be construed as local (close by) at a different scale of measurement. This helps us make sense of the process of resheathing: the activity of an entity or process at one scale that imposes an order upon entities or processes at a lower scale. Resheathing further illuminates the core proposal—that the universe is really and truly scale free.

Rather than repeat arguments I presented in earlier articles to the effect that Physics is nonreductive, what I will do in this chapter is focus upon an elegant formal methodology for handling the relation between macro and micro, called *renormalization*. Not only does it epitomize nonreductive Physics, illustrating the elegant nonreductive attitude the formal method takes in relation to quantities, the procedure is capable of making its formalism itself (in this instance it is a set of transformations) speak about whether the systems they describe are scale free. The model itself makes scale freedom transparent. (This sort of transparency has a way of creating hard-to-analyze loops of analysis, and headaches for logicians. But there's nothing especially worrisome about it—in fact, it has a certain kind of beauty.)

This procedure is the image with which I wish to inspire readers to examine the work of physicists; it is the best way to memorialize the message of this book. Renormalization is the quintessential mathematics of the scale-free universe.

2. Critical Phenomena

Phase transitions are an everyday affair and absolutely essential to life on earth: water boils in a kettle, frost forms on a window, and ice cubes melt in a fizzy drink. Similarly, iron is converted from paramagnetic to ferromagnetic, and certain materials undergo phase changes to superconductivity or superfluidity (remember the Cooper pairing) under the right conditions, or as a result of human-engineered preparation processes.

A phase transition normally involves either the release or absorption of what is termed a *latent heat*: the ice cube melting, the water evaporating, the frost forming on the windowpane. The latent heat is either absorbed or given off as the material makes a phase transition at a critical temperature. (We can sometimes extract that emitted energy and put it to good use.) The emission or absorption of heat is an indication that the structure of the material is being radically altered, the constituents entering an entirely new state of organization—*order* is actually the technical term.

A phase transition is under almost all circumstances an all-or-nothing affair: below the transition temperature we have one phase, and above it another, while at the transition temperature both coexist. With a phase transition comes an abrupt change of, among other things, densities and specific heat capacities.

In ordinary circumstances, phase transitions involve an exchange of latent heat. But under special circumstances, the phase transition is called *continuous*: there is no exchange of latent heat. One can realize these favorable circumstances by adjusting the pressure just so and slowly tweaking the temperature inside a stout vessel. For example, one can convert water to steam continuously (i.e., without boiling it) at a temperature of 647K, when the density of the substance is at 0.323 grams per cubic centimeter. These magnitudes of temperature and density define the *critical point* of the water-steam transition. (Above the critical temperature, water and steam cease to be distinct phases—there is just one continuous phase of water without sharp boundaries.) What one does is first heat the water to a temperature greater than the critical temperature under a pressure that exceeds that exerted by steam at the critical point and then subsequently reduce the pressure to any required value but keep the temperature constant (which is accomplished by increasing the volume in a continuous fashion).

There are many interesting classes of macroproperties of substances undergoing continuous phase transition, which have been very well studied. Specific heat capacities and compressibilities are among the more familiar such macroproperties. The magnitudes of many of these macroproperties near the critical point depend upon the location of the critical point (namely, the critical temperature and the density)—they reference the critical point—even when the magnitude in question (heat capacity, for instance) varies with temperature and not just at the critical point. For example, the specific heat of water at a temperature T is a function of (among other things) the difference between T and the critical temperature of water.

The behavior of certain macroproperties, near the critical point, is often in the form of a power law relation with some constant that defines the critical point. A power law is a (functional) relation between the macroproperty in question and another quantity, in which the magnitude of the one (the former, in this case) stands in an exponential power relation to the magnitude of the other (the latter). Here's a simple visual of the meaning of that last sentence:

$$C \propto \left|T_c - T\right|^{-a}.$$

The equation says that the heat capacity (C) is inversely proportional to the difference between the current temperate (T) and the critical temperature (T_c), to the power of α.

And, of course, the exponent is not itself another quantity; it is, as it must be, a dimensionless number, very often less than the unit 1, and never a whole number. It is called a *critical exponent.*

Here's the punch line: the critical exponents arising in very different systems are the same—such very different transitions as the liquid-gas transition of xenon and the separation of a mixture of two organic chemicals. This is a remarkable fact—or at any rate, it seemed so originally. The fact that so many substances share critical exponents is called *universality*. For example, the critical exponents of fluids of very different chemical compositions at their liquid-gas critical point are the same. More amazingly, they are an exact match for the critical exponents of the ferromagnetic phase transition in single-axle magnets. So we put systems sharing critical exponents in the same *universality class*. The features of systems that put them in the same universality class seem to be independent of the micro characters.

Once we have the classes, we can come up with a *renormalization group* (RG) that captures some interesting commonalities among the members. Any two systems in the same universality class will consequently have the same *dimensionality* (typically the number of dimensions recognized by the model as displaying interconnected order) as well as the same *order parameters*. An order parameter is a measure of the degree of order in a system; the extreme values are 0 for total disorder and 1 for complete order, and the value is often assigned quite arbitrarily, depending on the kind of system. For example, the order parameter for a substance between its liquid and vapor phases is defined as the difference in densities between the liquid and vapor. Both the notion of dimensionality and order parameter are features of the renormalization group. We cannot define these independently of the group: we must wait to find out what they are.

But what is a renormalization group? What's the point of this theory? From here on out I will be giving the reader my conceptualization of the procedure. I will contrast it with other people's conceptualizations of renormalization when I'm finished. Let the reader be the judge of which "bridge" between Physics and ordinary folk concepts is the better one.

3. Renormalization (the Procedure)

An expanse of matter in a dynamical context is characterized fundamentally by a certain function called its *Hamiltonian.* This function is meant to capture the quantity and the distribution of its different energies in different forms. We encountered this concept in chapter 1 when we were talking about mechanics and the Hamiltonian formulation. Ideally, this Hamiltonian would code for all such information as one purports makes a difference to the motions of individual bodies that compose the substance being treated—for example, the effects of interaction forces among a liquid's molecules, all external force fields acting upon it, and confinements such as boundary walls. (The postal address of the laboratory—please note—gets no special mention.) Of course the ideal of encoding all such information is impossible to achieve when there are on the order of 10^{23} microscopic bodies acting as independent sources of action within such a system. We thus require an approximation (remember our discussion of mean fields in chapter 2).

Until the 1970s all theoretical calculations of critical exponents were made in one of three ways: (1) by solving an exact equation of motion (by guessing at a Hamiltonian, for example as a function of thermodynamic macroquantities taken as "primitive," if not explicitly as degrees of freedom) for its thermodynamic properties, within the context of a favored thermodynamic theory, (2) by direct numerical simulations of such an equation (using, for example, a Monte Carlo method), or (3) by tinkering with equations that provide good enough accommodations of data *outside* of the critical region. But in 1966, Leo Kadanoff articulated an idea that would grow in time into the theory of renormalization. (A precise quantitative analysis presented in support of it by Wilson and Kogut in 1974 provided it the required credentials; see also Wilson 1975.) The idea, very generally, is that instead of seeking to home in on the right Hamiltonian (or the right solution to the right equation of motion) we can instead seek to transform the problem of finding the critical exponent to *the problem of the search for a Hamiltonian*—quite a different problem indeed—in an abstract *space* of conjectured systems (described uniquely by their Hamiltonians). The problem will have a solution only if certain features of the critical behavior we are examining—the critical exponents, for example—truly are universal. An astonishing idea: a Gödelian self-referential move!

Renormalization theory begins with a space—not of Hamiltonians proper, but of *effective* Hamiltonians. *Effective* means we're approximating (remember that we *still* can't write down Hamiltonians for systems with 10^{23} interacting bodies, even in the interest of hypotheticals). An effective Hamiltonian in the typical RG theory is a calculation of energies based upon a choice of a statistical distribution function for some variable connected with motion. It is, essentially, the "energy we would expect if" motion (for example) were distributed according to a particular recipe within the allowed range. The space of effective Hamiltonians is then a space of as many dimensions as there are factors to be taken account of when these effective Hamiltonians are articulated: we can think of it as the space of all possible system configurations for a certain very general sort of system.

A prefatory remark before proceeding. The sort of renormalization procedure I will be describing from here onward is concerned with the occupation of space by entities whose bulk properties we are interested in characterizing. The founding idea is to characterize the substance thereby constituted in terms of the way it occupies the space it has available. One can spell out renormalization in other ways as well, and this leaves exciting new avenues for further philosophical investigation. The tricky part of mounting a viable renormalization theory for a particular substance is to identify, for each space of Hamiltonians, (1) a blocking procedure (related to occupation of space) and (2) a renormalization transformation or *flow*, both of which will eventually have something to do with dimensionality and order parameter, if all goes well. Here's what this comes to.

1. The blocking procedure is an operation performed upon a lattice and depends upon the sort of symmetry manifested by the sites in that lattice. You may ask: What does a lattice have to do with the critical behavior of, say, a fluid? The lattice is the model of spatial occupation by (for example) fluid molecules. And the blocking procedure is a way of *replacing* one occupation scheme with another—by replacing groups of atoms thought of as "one" with another where the "one" is represented atomically (aerially, so to speak, as though from different geographical or scale perspective). (Figures 8.1–8.3 present visually several examples of blocking procedures.)

Renormalization of space is thus strategically a linked model (recall the taxonomy of chapter 1). I will not venture to speak on whether renormalization

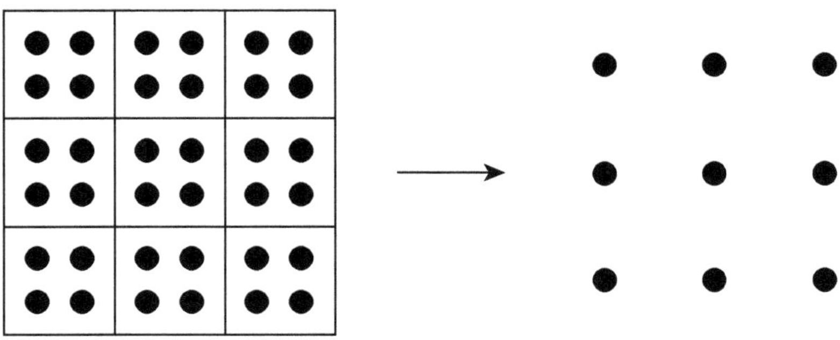

Figure 8.1 Renormalization of a square lattice. The linear dimensions of the lattice on the right must be shrunk by a factor of 2 to render it similar to the original. The final lattice has fewer sites than the original by a factor of 4.

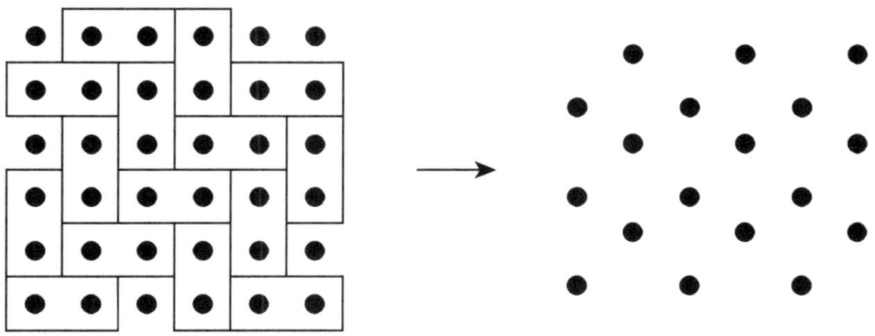

Figure 8.2 An alternative blocking scheme for renormalization of a square lattice. Here the linear dimensions of the lattice on the right must be shrunk by a factor of square root of 2 to render it similar to the original. The final lattice has fewer sites than the original by a factor of 2.

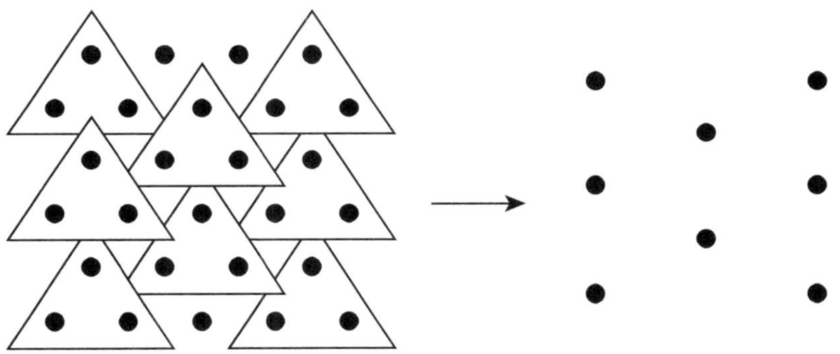

Figure 8.3 Renormalization of a triangular lattice. The dimensions of the lattice on the right must be shrunk by a factor of the square root of 3 to render it similar to the original. The final lattice has fewer sites than the original by a factor of 3.

procedures are all instances of linked modeling; there are too many varieties of renormalization to support facile generalizations.

2. The renormalization transformation, on the other hand, is a way of replacing ("flowing") by discrete jumps from one Hamiltonian representing the original occupation scheme, to the next occupation scheme (with potentially different occupants as well as different configurations of occupants) got by iterating the blocking procedure one more step. Therefore we can think of the renormalization flow as a *dual* or *image*, in the space of Hamiltonians, of the iteration of the blocking process.

Features of a flow—if we've selected a "sensible" one, which is to say, one that will get us results that are intelligible in these terms—will include attractive fixed points, repulsive fixed points, and saddle (mixed) fixed points (see figure 8.4).

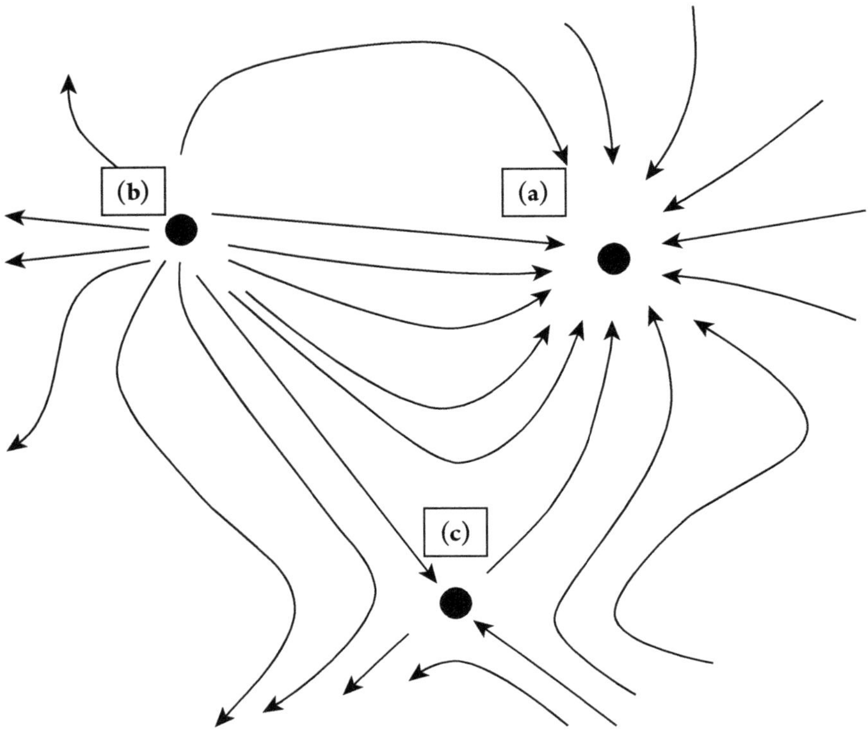

Figure 8.4 Illustration of the lines of flow about the three types of fixed points: (a) attractive fixed point; (b) repulsive fixed point; and (c) saddle point

Some of these fixed points—namely, the saddle points—will qualify as critical fixed points. And various lines of flow around critical points will be identifiable as *critical surfaces*—surfaces where critical behavior takes place.

If we achieve a picture of this sort, with a chosen RG, we will have a story to tell. Otherwise, back to the drawing board: pick another RG. The goal is to conceptualize flow in a suitable space of (effective) Hamiltonians that will allow us to tell a story about fixed points, in something like dynamical terms. But we are not speaking of *real systems in time* when we use the terminology of dynamics. That interpretation is inapt in this context, because what we're doing is transitioning between *different kinds of systems*—via flowing between different articulations or characterizations of Hamiltonians. What does flow represent? It represents some connection between one system and the next one in the flow: it is a connection by similarity.

It is very important to emphasize that RG is therefore making connections of similarity or as-if relations between system characterizations. For instance, an RG that produces the flow in figure 8.1 is articulating the idea that the block of four components is behaving as if it were one. This is on a par with the statement that a Cooper pair of fermions is behaving as if it were a single boson. The theory of renormalization makes elegant, systematic connections of this sort among abstractly characterized systems, arranging them in terms of these similarity relationships according to their *scale* relations to one another. Each transformation in the flow takes a step toward enlarging the "reach" of local conditions, thus expanding the scale of the "action" it is considering: in the beginning, we have action transpiring on a very small (micro) scale, and by the time we are finished "flowing," we are at a very high scale of action indeed, and locality (of the sort that physicists typically have in mind) has been left somewhere in between—maybe after the first step or two. The theory begins with characterizations of systems as linked. *Then* it proceeds to speak of their scaling relations, and in that way *also* speaks of them as more and more synced! In the terms we introduced in chapter 1, the theory of renormalization is indeed about ways that the linked becomes the synced. I thus see the theory of renormalization as making the philosophical point that we were trying (so inelegantly by comparison) to make in chapter 1: that linked models can have the same global features as synced ones, when the information "seepage" parameter is set high enough. (Admittedly, we made that point in chapter 1 without mathematical props of any kind, so maybe that was not such an inelegant formulation of the idea after all.)

Critical exponents can then be calculated from an RG model, as features of the location of and flow upon and across the critical surface near the critical fixed point. Critical exponents, on this analysis, come out as functions also of the parameters used in the blocking procedure (the order parameters and dimensionality, to which I promised to return).

Fabulously, RG never solves a single differential equation of motion. Indeed, how matters progress over time in any given system during any span of time is never as such a point of concern, for it is never the route by means of which to arrive at analysis of the behavior—in this case, the universality critical behavior—for which illumination is sought. The story we get from RG is entirely about taxonomies of order!

4. Explaining Universality

What explains universality? The single-scale universe model must cite the activity of the microscopic in interactions—perhaps the behavior resulting from force interactions among the microconstituents. But there is a problem—perhaps it deserves the name of a paradox: for on the one hand they must cite the impossibly many interactions at the micro or nano scales—the activities that by their lights *must* be responsible for the very existence of phase and phase transition, and on the other hand, they have got to admit that the details of these interactions cannot play much if any role in determining the critical exponents, since the critical features stay the same across different substances within the universality class, while the forces and the details of to-ings and fro-ings among constituents change radically from one member of the class to the next.

Here's an alternative proposal: macro features explain critical behavior. What macro features? They are the ones that RG opaquely refers to by the labels of order parameters and other technical what-nots. These may not be features of the substances that we are already familiar with and ordinarily measure, but they are features all the same that we might be able to learn to measure. I have a brilliant idea: why don't we go looking for them? We don't know what they are or how to measure them *yet*, but if we look to the RG that generates them for clues, and if we conduct measurements of many features in all the substances of the relevant universality class, and if furthermore we find relevant correlations, surely that will be a triumph for the scale-free universe model.

I deliberately avoided in the last section mentioning a certain singular fact: one very important feature of renormalization theory is that it provides no prescriptions or formulas in advance for helping us identify the blocking scheme. We don't have any recipes for passage from a micro description of the molecules involved to the "correct" renormalization group. We have no guidance except the list of members of a given universality class, self-admitted (as it were). So identifying the blocking scheme properly is part of the "art" of the method. But putting it this way functions to soften a blow: there is no formula for identifying the RG for any universality class—none whatever. This is an important point that we will put to good use presently.

For those of us who have embraced the scale-free universe, renormalization theory contains the seeds whereby to cultivate, in a most elegant way, a rich array of distinctions among macro categories that will enable the variety of macro categories (in Sokolowski's words) to stand forth for what they are. We can recruit this interpretation of renormalization in the service of the thesis that we need no longer understand macro categories in terms of a single unilluminating decipherment—in terms simply of the phrase "founded upon the micro." To us, renormalization can be interpreted as denying the need for this proposed decipherment: it liberates the macro. We can now go looking for each macro character, shared by the different members of a universality class. Surely they are waiting to be discovered.

5. Returns to Single Scale

Renormalization is one in a long series of chapters in the recurring saga of competitions between reductive and nonreductive physics. (This is a saga I first described in Thalos 2006.) Had the practitioners of renormalization declared openly a scale-free model of the universe, such as I am now suggesting we might need to do (in the interests of philosophical honesty) on their behalf, and had they separated from the reductionists and founded a branch of mechanics all their own, they would have diminished considerably the chances of being taken as seriously as they are (now). They have achieved their scientific credentials on the merits of looking upon their founding problems through the lens of a "paradox"—to the effect that somehow certain details of microstructure have to be overlooked even though microstructure must, ultimately, be given the credit for the behavior to be explained.

I was echoing statistical physicists when I called the problem of explaining universality a "paradox." And numerous practitioners of renormalization theory (maybe even the preponderance) to this day view their ideas as friendly to reductionist instincts. For example Binney and coauthors write, "there is a paradox here [!]: on the one hand it is clear that inter-atomic forces are responsible for the very existence of a phase transition; on the other hand the details of these forces cannot play any role in determining the critical exponents, since these stay the same when the atoms, and therefore the forces, change" (Binney et al. 1992, 21). Thus, ironically, many physicists see themselves as having to solve the paradox of holding on to apparently inconsistent views, when just possibly the simple resolution of their dilemma is to acknowledge the inconsistency and to cast the single-scale model of the universe adrift. The philosophical underpinnings of renormalization are most apparent upon removal of the false shroud of paradox. The "paradox" gambit is easy to play: it can be easily mistaken for the idea that we are simply "postponing" further analysis. Indeed this was fortunate in the beginning: it afforded new ideas entry in the guise of old. The contortions to keep the illusion of the sheep from dissolving into the reality of the wolf have been positively Herculean. My task here is to show that the reductionist credo need no longer command blind adherence because we are already proceeding in directions that defy it.

A certain proposition has emerged lately (speaking of contortions) in the work of the philosopher of physics Robert Batterman. He believes we can understand renormalization in line with the single-scale model of the universe credo. What explains universality, according to him, is the fact that the substances in a universality class share a certain select subset of microstructure—or perhaps better to call it a summation of such features—rather than all the many fine features of constituents and their interactions. With each step in an RG flow we "subtract" features. He proposes that we might cite such concepts as dimensionality and order parameters to explain what the different microstructures have in common, in virtue of which they give rise to the same macro behaviors. He thus defends the proposition that renormalization is citing nothing but a selection of the features of microstructure, which results from *subtraction* of detail.

Is this proposition worthy of our allegiance?

Khinchin was perhaps the first physicist to give expression to the "subtraction" proposition to which Batterman lends his support. Khinchin had the notion that the bulk properties of matter were derived from features of its microscopic interactions that are shared with other substances—common traits. He wrote:

> Statistical mechanics has for its purpose the construction of a special physical theory which should represent a theoretical basis for some parts of physics (in the first place, for thermodynamics) using as few special hypotheses as possible. More precisely, statistical mechanics considers every kind of matter as a certain mechanical system and tries to deduce the general physical (in particular, thermodynamic) laws governing the behavior of this matter from the most general properties of mechanical systems. (Khinchin 1949, 7)

By this idea of ascending to the highest level of generality possible, Khinchin sought to explain why Willard Gibbs's phase-average method of calculating thermodynamic constants succeeds.[1] The program Khinchin inaugurated has achieved a number of successes.[2] Khinchin's idea was to *assume* (fantastically!) that interactions between particles of a substance are negligible for purposes of calculating thermodynamic quantities, which are consequently written down as sum-functions of the right sort (thereby giving mathematical muscle to the assumption that interactions between individual particles don't amount to anything), and then to prove (by judicious applications of the Central Limit Theorem of probability theory, which sets down the conditions under which we can expect a normal statistical distribution) that time-averages for these functions approach the corresponding phase averages as the number of particles of the system goes to infinity.

Some very powerful criticisms have been made of the Khinchin program,[3] the most important for our purposes being that the assumptions for application of the Central Limit Theorem fail particularly (even spectacularly) in the critical regime. True: Khinchin did not aspire to explain critical behavior—he hoped merely (!) to account for classical, static equilibrium behavior. But one would expect that a theory that purports to explain via appealing to universal features

[1] Batterman (1998) elaborates the point and sets the Khinchin program in historical perspective as an alternative to ergodic theory. I think Khinchin's idea is not a good interpretation of Gibbs's program and set down my reasons in Thalos (2006).

[2] See, for example, extensions of this work to that of Sinai (1982 and 1992).

[3] Batterman (1998) gives the details and responses besides.

of mechanics would in point of fact be better placed to account for critical features than for equilibrium behavior, since this is precisely where "micro details don't matter." But it could not possibly do so, since its assumptions fail manifestly in these regions. I will refer to Khinchin's idea—to the effect that solution of numerous problems to do both with equilibrium and out-of-equilibrium thermodynamic behavior can be effected by rising to a high level of generality, thereby vindicating the idea that details don't matter—as *Khinchin's wager*. So I shall take the other side of this wager: handling macro phenomena by means of a "subtraction" strategy is misguided. Subtraction by itself is insufficient. Addition too is needful. How will this rivalry be played out?

Batterman (1998 and 2000) proposes to win Khinchin's wager by appealing to renormalization theory. He suggests that the renormalization procedure, which I outlined in the previous section, achieves just the sort of maneuver Khinchin urged: it achieves the elimination of microscopic detail, while bringing into sharper focus the most general features of a many-body system's physics. And it can do so for every regime of behaviors: it is not limited to treating just the classical regime, but can take account also of critical phenomena. Batterman's characterization of renormalization theory is as follows.

One introduces a transformation on this space (of Hamiltonians) that changes an initial physical Hamiltonian describing a real system into another Hamiltonian in the space. The transformation preserves, to some extent, the form of the original Hamiltonian so that when the thermodynamic parameters are properly adjusted (renormalized), the new renormalized Hamiltonian describes a system exhibiting similar behavior. Most importantly, however, the transformation effects a reduction in the number of coupled components or degrees of freedom within the correlation length (Batterman 2000, 127).

He then concludes, with heavy emphasis, that "the renormalization group transformation eliminates degrees of freedom (microscopic details) which are inessential or irrelevant for characterizing the system's behavior at criticality" (127), in effect, that renormalization theory vindicates the subtraction proposition, because "it is a method for extracting just those features of systems, viewed macroscopically, which are stable under perturbation of their microscopic details" (129).

But what could elimination of degrees of freedom mean for someone who adheres to the single-scale universe credo and advocates instead the subtraction proposition? What does it mean for Batterman? He does not tell us in so many words. His language suggests that all he means by the form of words is that the variables in the equations we must subsequently handle, at any given stage of renormalization, are fewer than at the stage previous. This meaning would make no connection whatever with degrees of freedom in our sense. In fact, this meaning has no metaphysical significance whatsoever.

It is illuminating, perhaps, to follow Batterman's argument from the top (or, rather, in the appendix to Batterman 2000). To do so we must begin by defining the notion of a *correlation length*, roughly, as the typical spatial distance over which the behavior of one microscopic quantity (here Batterman appears to equate "microscopic quantity" with "degree of freedom") can be correlated with the behavior of another. In essence, correlation length is the distance one has to go from a given particle before one can be fairly certain that its effects can safely be ignored for purposes at hand. Think of a lattice of spins. Above the critical temperature, and away from external magnetic fields, neighboring spins in the lattice interact in a way that tends to align them parallel to each other. This produces ferromagnetism. As the temperature is increased, however, thermal energy can randomize the spins, so that ferromagnetism can be lost. But as temperature is reduced toward the critical temperature, the spatial extent of blocks with correlated spins will increase (reaching "infinity" as we reach the critical surface). Think of correlation length as the extent of blocks with correlated spin.

Now renormalization, claims Batterman, is a process that preserves certain features of a lattice but refers to a trajectory along which there are variations in correlation length. Repeated application of the renormalization transformation results in "a sequence of Hamiltonians all of which describe systems with the same lattice spacing, but where the correlation length gets smaller and smaller with each iteration. As a result, the new Hamiltonians describe systems with fewer and fewer coupled degrees of freedom within their correlation length" (2000, 139). Movement along this trajectory, it would appear, is analogous to, or perhaps even identical with, taking a single system and adjusting its temperature—although Batterman himself is careful not to say that we have the same system throughout. The language in textbooks on renormalization utilizes the term "physical trajectory" for the trajectory Batterman describes. The problem is that this couldn't possibly be interpreted as a physical trajectory in the ordinary sense, as a renormalization transformation is officially a change of system, because it is officially a change of microstructure.

What Batterman seems to be claiming is this: if we can begin at different points in the space of Hamiltonians, and through application of the same renormalization procedure reach the same (critical) destination point, this shows that the beginning points that correspond to different substances will be possessed of the same critical behavior, and that this explains the universality we seek to explain.

But how? Where, precisely, is the explanation? Where, precisely, is the winning of Khinchin's wager? What, exactly, corresponds to the so-called *elimination of degrees of freedom*? After all, we are "flowing" from one system or sort of system to another. The referents of Batterman's analysis elude us. What is the *meaning* of the renormalization procedure—renormalization flow—that allows us to reach

Batterman's favored conclusion? All we are told is that renormalization flow emanating from different origination points can reach the same destination point. But what is the meaning of a destination point on the renormalization flow graph or the route taken to it? (On the interpretation I presented above, each point in the space—and so each point in the flow graph—represents a different type of system entirely, one with different degrees of freedom; the "destination" might be a system with very few degrees of freedom, something on a very high scale and representing a high degree of integration—like a rigid rod.)

The trouble we are having in understanding Batterman's position is this: once a renormalization transformation is applied to a Hamiltonian, we can no longer say of the result that it corresponds exactly to the original system or substance—or even anything that resembles it in features like its phase. So it is impossible to credit the interpretation set down by Batterman as a triumph vis-à-vis Khinchin's wager.

Alternatively, the proponent of a single-scale universe might wish to challenge the idea that there is such a thing as true universality—to gainsay all claims that universality classes do indeed exist as a reality, instead of as a purely convenient fiction. True, we seem to witness abrupt transitions of phase, but the true reductionist might well challenge the very reality of phase transition. There are only smooth transitions in the micro world. When we stoop to trade in macro concepts, we are simply dealing in approximation methods. Commerce in such ideas is literally meaningless, literally without referents; we engage in it only in the name of complexity management. There is nothing literally true to explain. There is nothing for renormalization theory to be accurately tracking, howsoever elegantly it proceeds.

This challenge will, obviously, come at a price. It challenges the way we are naturally inclined to report observations. If we accept the reductionist credo, we can no longer regard as absolutely undeniable the discontinuities that our observations would insist upon. When steam condenses on the windowpane, there is apparently in the instant when the steam's temperature falls sufficiently low to enable (at the given pressure) the transition to occur, an instantaneous change of the bulk properties of the substance H_2O undergoing the change. Latent heat is instantaneously released—or anyway, so it seems. It would seem that the boundaries within which these transformations occur are absolutely rigid: they are constants for that substance, which (according to the so-called *scaling hypothesis*) are related to one another, as well as identical with the constants for every (completely different) substance in the same universality class. And so what seems to be absolutely undeniable is the fact that the constants—and thus the quantities associated with them—belong to a macro structure that is shared among substances that are as different from one another, when it comes to the microscopic, as substances can possibly be. The single-scale credo requires us to deny the absolutely undeniable.

That price is too high. The open-minded might, in the face of such an unreasonable demand, simply regard universality as evidence of the scale-free universe, some measure of corroboration for the fact that the universe is active at multiple scales, and that systems in the same universality class share something macro that importantly governs their behaviors in the critical regimes. The bulk properties of a substance do not follow from its microstructure, because nothing in the way of micro is shared between the different substances that belong in the same universality class. Furthermore, the same suites of macro properties are not prohibited from being shared by substances with different microstructure. Of course this fact—if we choose to allege it is a fact—is extraordinarily hard to establish decisively. For it is a negative thesis: that there is nothing in the way of micro structure shared among the substances in the same universality class. So we do not have proof positive of the scale-free universe. Nevertheless the data is certainly speaking to us—very loudly—in those terms.

6. The Liberation of Macro

What lesson might we take away? We might conclude that the dynamics of constituents of bulk matter—as such—has little or nothing to do with the macro features of its behavior. We might conclude, in other words, that the bulk features of a macro body owe nothing of substance to the micro interactions among the constituents of the matter of which it is comprised—that its macro features are independent (in the sense defined in the previous chapter) and are properly governed by laws that are not dreamed of in reductive mechanics. Hence, we must confer degrees of freedom upon the macro and regard such features of the bulk as its temperature as the appropriate ones in terms of which to give a "rock bottom" analysis of certain phenomena—where "rock bottom" no longer means that it refers to the micro. The nontrivial work in science will become—as it has always been, dogmas notwithstanding—to identify the quantity or quantities that deserve the name of a degree of freedom, and then to proceed with articulating a theory according to which this quantity plays a key role.

This conclusion will become a launching point onto the sort of philosophical analysis of real-life physics that has appeared multiple times in the chapters of this book, a sort of analysis not available when we were restricted to micro degrees of freedom exclusively.

Of course we could never *prove* this negative judgment that there is no dependence (in the sense defined in chapter 7) of macro upon micro. After all, it's precisely because the mathematics of dealing with something on the order of 10^{23} variables is impracticable that folks have had to resort to different means of analysis. But the very fact that one can do so, and be satisfied with the results, *suggests we do not have to lay claim to the alleged fact that these results would follow also from an analysis that took a pathway through the micro as the most fundamental reality on which all rests, subtracting away where necessary whatever is irrelevant.* It is precisely this purported fact—that the micro is the most fundamental reality on which the macro rests—to which many folks dearly love to appeal, so as to stay on the right side of the reductionist credo. I offer two remarks in conclusion. First, the appeal to the micro is a promissory note that has never and could scarcely be made good. Second, it is quite inconsistent, or at least out of consonance, with much of what we can legitimately say about the procedure of renormalization, because renormalization says nothing at all about the dynamics of systems—there are no dynamical principles in the theory whatsoever.

Finally, I would remind the reader once more of the singular feature of renormalization theories of the critical phenomena: they provide no prescriptions or formulas whatsoever to help the theorist identify the blocking scheme. There are no recipes for how to construct an RG for a given universality class. But if we are looking for commonalities among microstructures, surely there would be. After all, that's the presumption. Reductionists can never concede that, in theory, coming up with the right blocking procedure is an "art." A recipe for blocking procedures would go a long way toward winning them Khinchin's wager.

Until such time as a general recipe for RG procedures has been produced from some sort of "first principle" that govern microquantities, renormalization theory stands as a liberator of the macro. Liberation of the macro removes the liberated into a new realm of order, a new filter, but without denying its relations (of interlocking) to other realms. This is what unification consists in. Thus we see that the unification of Physics consists, not in the elimination of macro (as a slave to micro), but in its formation of new realms of order that interlock with micro realms as well: unity without hierarchy, a marriage of equals. This view throws light, and not cold water, on profound questions about order and how the universe, in its multiple scales, holds together. This is therefore a unity worthy of Science.

Conclusion

Toward a More Truth-Illuminating Metaphysics

> In both science and engineering, the study of "systems" is an increasingly popular activity. Its popularity is more a response to a pressing need for synthesizing and analyzing complexity than it is to any large development of a body of knowledge and technique for dealing with complexity. If this popularity is to be more than a fad, necessity will have to mother invention and provide substance to go with the name.
>
> —Herbert Simon, *"The Architecture of Complexity"* (1962)

We have taken some baby steps of a long journey toward a faraway destination. Along the way we have (I hope) appreciated the fact that the single-scale model of the universe, even when dressed up in emergentist regalia, is reductionism in sheep's clothing. It takes some knowledge of the wolf to be able to detect one "under cover." Reductionism, as we now see, has a constricting and altogether negative effect upon the creation and pursuit of novel scientific ideas. To fully appreciate the constrictions of reductionism upon scientific thought we have to acquaint ourselves with its antithesis, so to gain a view of the way of sheep as compared with that of wolves. And this is why (among other reasons) we require the model of scale-freedom that we have been developing in this book—it is to help us keep sheep in mind. The scale-free model of the universe can also serve as the definitive antireductionist manifesto.

The scale-free model is furthermore evocative of questions that have simply never been articulated before. The model is thus pregnant with possibilities for future inquiry in natural history and metaphysics. And it is equally pregnant with puzzles to reinvigorate philosophy. For example: if resheathing is the way by which new things come into being, and if resheathing is ordinarily the process whereby entities at a certain scale act upon entities at a slightly lower scale, then how does the "first" resheathing take place? (A potential answer—drawing on things we've already articulated in this book—is that certain processes, like Cooper pairing between electrons, are *spontaneous* at a certain temperature so long as the density of matter is high enough; and it might therefore explain why the universe had to begin spatially small. Still, this requires a great deal more working out.

What Philosophy would need to do to fill out this proposal is to work out a theory of spontaneous processes—and what that amounts to in a scale-free universe.)

It is truly dogmatic to insist on Physics as the Master Science and designated answerer of all questions pertaining to the true activity of the universe. And it is dogmatic in the same measure to insist that all the real "action" in the universe happens at the minutest scales. Indeed, they are one and the same ill-considered dogma. The intellectual community should demand open-minded study of the question of how scales of behavior are related to one another. So we require a true science of systems—a science that is at one and the same time a theory of metaphysics. As Herbert Simon insisted, we require a "substance to go with the name." And this book has put forward a sustained plea for substance.

For a few years now, the Santa Fe Institute has been stepping up to the plate. The SFI has brought scientific focus upon a number of the questions I've arrayed in this book. But very few philosophers have taken note of the SFI's efforts as bearing on the full array of metaphysical questions. Philosophers, especially philosophers of mind, have resisted the notion that metaphysical questions of this large magnitude can be addressed by empirical inquiries. Metaphysical questions of large magnitude, according to this line of reasoning, can only be handled a priori, in the armchair. And relatedly, in connection with issues regarding large networks and their large-scale properties, the prevailing philosophical opinion is that we need systems analysis because of the "complexity" created by many entities: there are far more "basic" entities, more bodies, more particles, and hence more quantities than a mathematical model founded on "fundamental" physical laws can handle. This makes things mathematically more "complicated." And by "complex" and "complicated," philosophers just mean "messy." Things get many orders of magnitude more messy as we broaden the scope of concern to topics outside of two-body physics. Thus, the necessity for a less fastidious style of analysis is a matter of practicality, not a matter of principle (an "applied" rather than a theoretical matter). Systems analysis is thus, according to the prevalent view, a concession to human frailties. We stoop to systems analysis because we must navigate the world with finite mental resources Complexity is a practical obstacle, and thus an occasion for being practical in making concessions—for yielding to a less rigorous form of analysis. Complexity, in this sense, calls for nothing more than simplification for practicality's sake.

I argued in chapter 3 that this point must be strenuously repelled, if for no other reason than that this conception of complexity is mistaken. For it is an unsubstantiated dogma—embraced of course a priori, since simple empirical studies would contradict it immediately—that the facts about networks or aggregates are already contained in the facts about the node occupiers or aggregants. It is a further unsubstantiated dogma that the laws governing behavior of aggregants and node occupiers are the laws that govern all behavior, without exception. And so we must reply to Systems detractors as follows: we need Systems, *not* because the world is so messy, but rather *in spite* of this fact. For the world at larger scales is not as messy as the numbers of entities inhabiting the smaller scales might aprioristically suggest. This is, in itself, a metaphysical truth of high order.

The theory of Systems—a theory that rises to the level of metaphysics—does not oversimplify what is *genuinely* much more messy. Rather, as we have seen here, Systems analysis is capable of taking advantage, in an elegant way, of true reductions in that complexity—*the true reductions made by the universe itself in the degrees of freedom*. These reductions are manifestly *not* due to human oversimplifications; they are due instead to high-order truths regarding the stability of aggregates, sheathed and networked in ways that are inconspicuous to those with eyes only for the micro.

To those with open eyes, the world appears—and positively not deceptively—to possess many layers of fundamental structure at many scales.[1] And this is what Systems analysis (with a capital *S*) seeks to describe. Accordingly, we are intellectually bound to reply to the critic: we require Systems analysis because there is more order—and correspondingly *less* messiness—in heaven and earth than is dreamt of in the aprioristic philosophy of atomism. Describing that order requires theories that are capable of making reference to higher-scale order, to higher-scale properties, to relations between facts at different scales, and ultimately to theories of how entities at different scales interact. An account of this order and how it is attainable, in metaphysical rather than in purely domain-specific terms, is a far firmer foundation than we have had to date for a theory of science that aims to accommodate each special science as autonomous.

Physics, as the previous chapter demonstrated, can in fact be viewed as a microcosm of all of science. Many of the philosophical problems that concern science generally, also concern a range of physical sciences specifically. And so scrutiny of problems facing Physics itself has sparked many of the most exciting contemporary developments in conceptualizing the interaction between entities and the aggregates in which they conspire, their aggregate behavior, and how, through participating in aggregate behavior, they manage to exercise power over one another. Studying relations between the various subdisciplines within Physics, and theories that cross these boundaries, is thus bound to reveal how Science more widely can enjoy a unity as a result of overlapping areas of concern among the sciences.

[1] Simon gestured at this idea almost fifty years ago: "How complex or simple a structure is depends critically upon the way in which we describe it. Most of the complex structures found in the world are enormously redundant, and we can use this redundancy to simplify their description. But to use it, to achieve the simplification, we must find the right representation" (1962, 481).

One result of our examinations has been to reveal that aprioristic metaphysics cannot dictate boundaries between disciplines, as it has been wont to do. Aprioristic metaphysics cannot, for instance, dictate a boundary between sciences of the physical and sciences of the mind—the boundary with which it has been the most consumed. Aprioristic metaphysics has no light to shed on these matters. Descartes's watershed argument in favor of dividing the physical from everything else has done major damage to the prospects of a unified science. We are to this day wrestling with the aftermath of Descartes's philosophical accomplishment. And appreciation of this fact might, if we are especially careful with the insight, help us to reinvent Philosophy.

1. Resheathing and the Glue of the Universe

The most important philosophical innovation that this book will have wrought, if it's successful, is the notion of *resheathing*, supported by the conceptions revolving around the idea of *dependence* that were laid out in chapter 7. The notion of resheathing makes it possible to talk about a scale-free universe, in which action of every sort (but especially resheathing) occurs at every scale of measurement. The cement of the universe—what makes an entire universe, comprised of many filters at different scales, into a single unity—is a single budget of dependent and independent quantities, united by a logic of leading and following. The laws of nature join together in a marriage of all scales to provide a set of tools for managing this single budget of leaders and followers. Hence, the unity of the sciences.

These conceptions can reinvigorate metaphysics.

2. Philosophy Reinvented

> We have painted ourselves into a corner of irrelevance so completely that at times I wonder whether most philosophical work is even very interesting to other philosophers.
>
> —Lee McIntyre, *"Making Philosophy Matter—or Else"* (2011)

Academic Philosophy, especially the "core" discipline of metaphysics, nowadays suffers the unenviable reputation of being occupied in idle or insubstantial employments of the intellect, in pursuit of verbal games and puzzles whose solutions are largely insignificant for human life in general and for the vast majority of academic concerns in particular. And it is quite true that many contemporary philosophers working in traditional areas of metaphysics and epistemology are completely disengaged from developments in the sciences; the conviction that science, or at any rate the specific findings or theories therein found, are profoundly irrelevant to the problems of Philosophy. But we must not forget that in the beginning Philosophy, drew all intellectual activities into a single unifying framework. Because it did so, Philosophy was at that time able to effect transformation upon transformation of the intellectual landscape. A single idea at a given node could dramatically alter the entire landscape—as for example Descartes's ideas did. Because the sciences were unified, Philosophy had the advantage of scale.

And today there is no question that, if human intellectual activities are to be drawn once again into a single unifying framework, it must occur through a philosophical innovation. If a unified academic enterprise will emerge, it will have to be because Philosophy has risen to the challenge of making it happen. What other discipline is in a position to bring about such a transformation?

Philosophy did not create the present balkanization of intellectual activity—at least not intentionally. Philosophers have historically ranged quite broadly in the space of ideas, never supposing that any idea or region of the space of ideas might be off limits to them. Once upon a time Philosophy played the role of an active coparticipant in the growth and development of Science; philosophers were also scientists and did not insist upon an exclusive preserve that belonged only to Philosophy. In the earliest days of Philosophy, Aristotle was the foremost biologist of his time—and famously held that Philosophy acted as a midwife to Science, by articulating discipline-founding concepts; in the modern period, William James was the father of modern Psychology, standing at the turn of the twentieth century with feet solidly planted in both Philosophy and Psychology.

In the mid-twentieth century everything changed. So that now, near the beginning of the twenty-first, Philosophy seems to have withdrawn from the frontiers of knowledge to an exclusively philosophical cache of topics and concepts. However and it has left behind a damaging legacy, not necessarily all of its own making: the conception of a one-scale universe. The task of this book has been to contribute to the dismantling of that legacy, and in the process to reinvent Philosophy within a reunified scientific field.

What does it take to reunite the community of inquiry? There are numerous ways that sociologists have developed to classify adhesives or bonds that hold groups together. From the very beginning of the discipline, sociologists have sought to taxonomize social organization, recognizing a great variety in the wild. For instance, the German sociologist Ferdinand Tönnies (1855–1936) is best remembered for a distinction between two types of social groups parallel to two purportedly basic forms of human will: the essential will, which is an organic or instinctive driving force, and arbitrary will, which is deliberative and goal-oriented. Accordingly Tönnies (1957) referred to associations that form or organize around essential will as *Gemeinschaft* (often translated as *community*), while those in which membership is organized or sustained by some instrumental and explicit goal were termed *Gesellschaft* (often translated as *civil society*). Tönnies held that *Gemeinschaft* is best or most perfectly exemplified by the family and *Gesellschaft* by the state.[2]

The *Gemeinschaft*/*Gesellschaft* distinction is clearly a fellow traveler of a more contemporary distinction between individualist and collectivist societies. Modern America lies on one extremum of this continuum—the individualist extremum—while Asian societies fall closer to the other end of the continuum. Younger societies, interestingly, with fewer ties to ancient civilizations, gravitate toward the individualist end of the scale. Numerous measures are now deployed to diagnose societies on the dimension of collectivism, with the diagnostic tools being statistical measures applied to aggregations of answers to a range of survey questions. Still, the philosopher and sociologist R. G. Collingwood's description (1927, 23) of the collectivist mind is hard to improve upon:

> The individual counted for nothing except as the member of his guild, his church, his monastic order, his feudal hierarchy. Within these institutions he found a place where he was wanted, work for him to do, a market for his wares. He could devote himself to fulfilling the duties assigned him by his station in that great organism within which he found himself lodged.

[2] More on these sorts of taxonomies can be found in Thalos (2008). Individuals in *Gemeinschaften*, according to Tönnies, are regulated by shared norms that dictate the appropriate behavior and responsibilities of members of the association, to each other and to the association at large (there is a "unity of the will"; Tönnies 1957, 22). *Gemeinschaften* are broadly characterized by a moderate division of labor, strong personal and family ties, racial and ethnic homogeneity, and relatively simple organizational structures. In such societies, because there is a strong collective sense of loyalty and belonging, codes of conduct require little or no external enforcement. *Gesellschaften*, by contrast, lack shared norms of conduct and are maintained instead by individual self-interest. They are associations in which, for the individual, the larger association never takes on more importance than individual self-interest. A modern business is alleged a good example of *Gesellschaften*, where the workers, managers, and owners may share very little by way of concerns, beliefs, and motivations. Unlike *Gemeinschaften*, *Gesellschaften* emphasize secondary relationships rather than familial or community ties, and there is generally less individual loyalty to the association as such. Social cohesion in *Gesellschaften* presumably derives from the glue of an elaborate division of labor, which results in a profound and typically insurmountable material dependency of individuals upon one another and the system of organizations under which their labors are coordinated.

The Renaissance, according to Collingwood and others, broke with this culture. It gave birth to modern individualism, expressed in "the freedom of discovering that one can leave one's ordained place and march out into the world without being struck dead by an offended God" (1927, 30–31).

Collingwood held that individualism is a newcomer and indeed that the individualist quest for freedom led to decoherence of certain activities of mind—the aesthetic, religious, and scientific modes—that originally cohered much more closely. And, according to Collingwood, God was indeed offended, for this freedom comes at the price of an internal conflict, which is the disease of modernity. The curse of modern individualism is, therefore, the deep cause of the miserable condition—this fragmentation—of modern consciousness. Alternatively, one may celebrate the proliferation of expertise, intellectual specializations, and disciplines—this multidimensional, even baroque, division of labor—both within and without the university. My hope is that the sciences can be unified through a mutual understanding that, through an elaborate division of labor, the sciences collaborate to throw light on a scale free universe.

Today Philosophy seems to be on the outside of center, looking in, its bonds to its closest intellectual kin in tatters. I will leave the examination of how that happened to others. Instead, I have been concerned about how the reunification might be effected. I have talked about how domains of inquiry are interconnected. I have proposed intellectual foundations for the view that the entire field of inquiry is best construed as an interlocking system of areas of coherent investigation. I have argued that indeed all inquiry is best set within the context of shared intellectual vision, and that Philosophy must be, as it has always been, the basis of any such vision. This doesn't mean that only philosophers are in a position to provide vision or view the entire field of human knowledge; what I mean is that it is inherently a philosophical exercise—open to all researchers—to view one's inquiry as part of a larger enterprise. This requires a conception of inquiry less in terms of division of labor than in terms of bringing together disparate tools and ministrations for inquiring into the same things. Still, there are many obstacles to this vision of the field of inquiry becoming a reality, not least of all the numerous artificial disciplinary, funding, and institutional boundaries. And one of the least recognized sources of the current condition of the academy is Philosophy's own current self-conception.

It is commonplace nowadays for philosophers to feel at home "clearing away conceptual confusions." While one must not underestimate the value of this task, it must be admitted that it is not an ambitious one. But my vision in this book has been that Philosophy does some of its best work by helping to raise the profile of radical questions, while of course continuing to do what it has done consistently well: develop and hone, with annotations and commentaries, a repository of arguments and argumentational methodologies, probes, and analytical instruments, bringing each to bear as transparently as possible upon sanitized application cases, specifically for evaluating the merits of posited questions and proposals for addressing them. It is this process that keeps the academy vibrant. It relentlessly breaks down artificial disciplinary and institutional boundaries and makes of inquiry a genuinely translocal activity—in all the best senses of the term.

By cultivating a vision of the organic unity of science and by centralizing in the process those open-ended, discipline-inspecific problems that Philosophy itself has grappled with in its long and sometimes illustrious history, Philosophy can further shape the radical questions that revitalize inquiry. We have already discussed how philosophical problems about the nature of scientific subject areas get reinvented, as well as profound metaphysical problems about fundamentality. Here are just two more important areas.

3. Determinism

> We may regard the present state of the universe as the effect of its past and the cause of its future. An intellect which at a certain moment would know all forces that set nature in motion, and all positions of all items of which nature is composed, if this intellect were also vast enough to submit these data to analysis, it would embrace in a single formula the movements of the greatest bodies of the universe and those of the tiniest atom; for such an intellect nothing would be uncertain and the future just like the past would be present before its eyes.
> —Pierre Simon Laplace, *A Philosophical Essay on Probabilities* (1814)

For several centuries now, ever since the French mathematician and astronomer Laplace conceived of what we now refer to as Laplace's demon, a being capable of foreseeing the future of the universe once completely apprised of its present condition, philosophers have taken this ability to predict the entire course of the future to be a scientific aspiration. Obviously Laplace's demon is a genuine possibility only if the proposition now called *determinism* is true:

> *Determinism*. Given a set of conditions at a certain time T_0, there is only one future compatible with it for all future moments $T > T_0$.

Determinism is the proposition that there is only one future compatible with the present. And so it has been thought that determinism is incompatible with the notion of *freedom* or *freedom of the will*, which requires that the future be genuinely *open*—that it be genuinely true (rather than merely felt to be so by someone poised to act) that more than one future is compatible with the present one. Contrary to this thought, *compatibilism* maintains that freedom is indeed compatible with determinism, so long as all the facts about an agent (including about the agent's mind or will) are also aspects of the present, and so features or facts that Laplace's demon must take into account when he runs the algorithm he uses to predict the future.

The debate on this topic has been stalled for some time, and all the activity on the issue for the preponderance of its history has been on the question of whether compatibilism is defensible.

The position we have been carving out in this book opens a new window onto a certain move in the space of alternatives—a move that, although already conceived of, has not been pursued in significant ways, perhaps because it has not seemed philosophically attractive. The move is this: that determinism is not a fundamental principle of scientific analysis. Even if it might be an aspiration of some scientists, it is not part of the logic of science, not the basis of the unity of science, and in general it does not lend any substance to a theory that it be deterministic—not even in Physics. It helps to have a theory of science—which we have been developing—to be able to make this statement. In the context of the present work, this is not merely an allegation; instead it rests on a substantial theory of science as a unified framework of interlocking inquiries. I will take just a few more paragraphs to articulate what the new opening amounts to.

The cachet of determinism has for some time lacked luster, it is true, at least since we have come to appreciate the probabilistic nature of quantum theory. (Indeed this was the source of Einstein's dissatisfaction with the theory he helped to articulate.) But while it might be true that quantum physicists are not bothered by the loss of determinism, this failure has not seemed to be an obvious boon to those with concerns about freedom. It is no better for a theory of freedom that the universe is not so much deterministic as stochastic. If the explanation for a failure of determinism is instead randomness or chance, the prospects for freedom are no better, and we are better off with compatibilism. The thought here is simply that, if we have to give up the understanding that the present and the future are linked by physical (or at any rate causal) connections, the prospects for a theory of free agency is even worse; because it doesn't help the cause of free agency to posit that present and future are connected by chance rather than some sterner stuff.

I initially presented determinism as a challenge to the notion of free agency, something that must be overcome, because free agency requires an open future. The compatibilist, by contrast, often enters the debate with the proposition that free agency is enhanced rather than diminished by the proposition of determinism; and we should instead give up the idea that free agency requires an open future. The compatibilist who sees things this way will *not* welcome argumentation that casts doubt on determinism.

The scale-free universe model prompts an evaluation of the thesis of determinism for a theory of free agency. Does determinism enhance or diminish the prospect of free agency? In particular, in the light of the idea that the universe is active at every scale, how would determinism enhance or detract from the prospects of free agency? To answer that question, we need to enlarge our understanding of what determinism demands in a scale-free universe. This is perhaps not so obvious as how to understand determinism in the context of a single-scale (micro) universe.

The fundamental bone of contention in the debate over determinism is, as I've sought to convince you, the importance of the open universe conception. If freedom requires an open universe, then determinism and freedom are incompatible; but if instead freedom requires or is in some other way enhanced by the truth of determinism, freedom requires a model of agency (in particular, human agency) that is compatible with determinism. It seems therefore that compatibilists are obliged to offer a theory of human action compatible with determinism. It might therefore serve them to inquire whether the scale-free model of the universe is a more hospitable one for their views. At the same time incompatibilists might find it easier to dispense with determinism if they embraced a scale-free model. It seems quite possible that the scale-free model enhances the prospects of freedom, whether or not compatibilism is correct! Still, this requires some further working out.

4. The Nature of Body

Philosophy has often given itself over to wonderment about consciousness and its place in nature: how it is a category apart from that of body, how its nature is elusive, and how it doesn't make sense in the context of a universe shaped by unintelligent evolution. Mind (according to Descartes) is active, while body is passive. Mind is expansive—it reaches out, it "grasps" and holds things within itself, as a mirror of nature, with concepts and language; all that body can do is yank and thrust. Still, as I mentioned several chapters ago body is really very puzzling indeed—at least in comparison with more intelligible things like waves and fields. (This is no small reason why particle physicists seek mightily—but not always successfully—to construe all particle-like things as manifestations of fields.) But body possesses marvelous features. It can localize itself within the most minute regions of space and time: it is the very epitome of discontinuity. With this ability, body marvelously renders the universe granular, at one and the same time breaking the numbing symmetry of sameness-in-all-directions from which the universe might at one point have suffered. Body has qualities of self-protection: it resists being altered, whereas other forms of existence—for instance, waves or fields—have none of this profound resilience. Waves and fields are like wisps of smoke: insubstantial and malleable, offering no resistance whatever, agreeable to being altered, but nonetheless indestructible for it. Body is stolid and true, but typically not permanent. Body is intelligibly counted and reidentified; fields and waves merge and lose their identities.

Fields and waves are easily understood, highly intelligible. Body is that which is truly foreign to thought. How is it that a body manages to collect and hold characters unto itself, however impermanently? And if we can figure that out, we then have to work out how something that is localized manages to interact with something insubstantial and "everywhere at once"—we'll need a theory of inter-interaction that makes sense of these categories.

A recent provocative line of thought has gone radically to the opposite point of the suggestion I am making here. Kathrin Koslicki (2008) proposes the contrary to the idea prevailing among particle physicists that bodies are really, covertly, manifestations of waves or fields—in other words, rather than take for granted that (disembodied) structure is the true simple-to-understand point of departure for analysis, instead it is the other way about: structure can figure as a part—analogous to material parts—of bodies.[3] Structure can enter into a mereological (part-whole) calculus in exactly the same way that material parts do. The logic of bodies is parthood, which is the logic of part-to-whole; and this logic is nowise foreign—it is the most simple of all. Consider the classic case of the statue and the lump of clay from which it is fashioned (Koslicki 2008, 181). Clay and statue are distinct (since each has features the other lacks). Still, they share the same material constitution; the lump is a mere proper part of the statue. And so it follows that there is another proper part of the statue that is immaterial—namely its *structure*. Here Koslicki is reviving an idea originated by Aristotle (hylomorphism). And it is a very nice contrast to the position physicists seem to be taking nowadays.

In the end, what is perhaps most difficult to figure out is not so much which is the more basic or more simple—whether body or structure—as how the two interact if, as seems to be the case, they belong to fundamentally different categories. This is truly a metaphysical puzzle worthy of considerably more attention than it has ever received. The perspective on characters we have outlined in this book, especially in the conception of resheathing, should help to invigorate this subject.

5. God, the Universe, and Nothing Else: The Elephant in the Room

A certain patient reader, having worked through this book to the end, might now lose patience entirely. I've come nearly to the end of the journey but said yet not a single word about a certain elephant in the room. That reader might have seen the

[3] Of course she would not put it this way. Hers is a work in analytical metaphysics, not in scientific metaphysics.

obvious applications of a scale-free model of the universe from the beginning—an application to the areas of spirituality and religion, or to certain areas of environmental philosophy. Obviously, if I have wrought a successful model of the universe as active at every scale, this model (either to its credit or to its disgrace) will be available for scaffolding or supporting each of two kinds of views. The first is advanced by people who believe in a certain *highest* scale of activity—as contrasted with a single and very low scale of activity. The second is advanced by people who insist on a scale of organization that comprises all of life—a kind of *Gaia* scale of activity. (In the latter category are the adherents to the so-called *Gaia Hypothesis*, which proposes that all organisms and their inorganic surroundings on Earth are closely integrated to form a single and self-regulating system that in virtue of being so integrated is alive.) The first view is one that some theists might find attractive; the second is attractive to spiritualists and certain environmental philosophers.

Each view outlined in the preceding paragraph can be made to accommodate a certain feature that its proponents might find desirable: it can be made hospitable to the idea that there is a limit to the extent that the scale of activity of the universe can grow—it can be made to accommodate a cap on the scale of activity of the universe. Obviously, this is so for reasons that have nothing whatsoever to do with science. I personally find both views unattractive. As I mentioned at the start, I am a physicalist; but as I also mentioned, physicalism—as the view that everything is physical—is not a very demanding religion.

It is not my intention to engage with this scale-cap view. I will simply discuss some of its implications.

First, if a view embraces everything I have said thus far, it must accept that the existence of this scale of activity for the universe is just as opportunistic or contingent as the existence of everything else. So the upper level of activity is no more fundamental or necessary than any others. Second, this high scale of activity cannot be the source of the universe, nor the source of the activity of the universe; if anything, it is beholden to the lesser forms of order or activity for its existence—at least, those into which it is importantly overlapping and linked. This is true at every other scale—every new scale of order will enter into existence by overlapping with others; so from its start every character at a new scale will be interlocked with other characters in a network of dependence relations. Finally (and this follows from the previous point), every character at this scale will be subject to being acted upon by characters within its filter—in particular, it will be subject to being acted upon by precisely the set of characters with which it is interlocked. Of course we cannot say which will be leaders and which followers in advance—in fact that's the whole message of the book. But it is important to note that a theist cannot insist that the characters of God will always be the leaders with just the metaphysics I have sketched here. A theist who is comfortable with this proposition will be an instance of a rare breed.

More importantly, a person who embraces the divine or the spiritual as a scale of the universe, in the sense we are using the term here, will be embracing the framework of science for the articulation of their pronouncements.

I welcome this framing of the claims of spiritualists and other religionists. It is a brave move, a breath of air in a currently airless arena of speech. Religionists in particular have often declined to move their doctrines into the scientific space of intellectual inquiry. They have refused to allow the scrutiny associated with scientific inquiry to fall upon their proclamations. And often in an attitude of graciousness, they have, like Pope Pius XII, simply declared the nonoverlappingness between Science and Religion—in a gesture of giving science its freedom from oversight. Pius was especially concerned that the doctrine of evolution might be thought to cast doubt on the religious view that God created the universe. Accordingly he simply declared, by fiat, that the conclusion is unreachable from scientific premises:

> The Teaching Authority of the Church does not forbid that, in conformity with the present state of human sciences and sacred theology, research and discussions, on the part of men experienced in both fields, take place with regard to the doctrine of evolution, in as far as it inquires into the origin of the human body as coming from pre-existent and living matter—for the Catholic faith obliges us to hold that souls are immediately created by God. (Pius 1950; Gould 1999, 78)

And the biologist Stephen Jay Gould quoted this passage with approval in support of his doctrine of NonOverlapping MAgisteria (NOMA)—essentially accepting (perhaps benightedly) that the pope has released biology from his oversight: "science and religion do not glower at each other . . . [but] interdigitate in patterns of complex fingering, and at every fractal scale of self-similarity" (Gould 1999, 65). According to Gould, they "interdigitate" but do not overlap. But overlap is the core assertion of the scale-free model of the universe.

Of course NOMA is subject to philosophical disputation. But if religionists were to embrace the scale-free model of the universe as articulated here and supplement it with the further hypothesis that the divine is another (perhaps the highest) scale of activity of the universe, the religionists will not be allowed to hide behind the curtain of NOMA vis-à-vis their other doctrines. All their pronouncements will come open to empirical scrutiny. As I see it, that day cannot come too soon.

6. Toward Truth

Everything I have said thus far rests on a premise: that the unity of the world (such as I have depicted it) automatically justifies requiring some corresponding unity in the sciences. The benefits are many. But a critic might now undertake to argue that all these benefits can be enjoyed even if one is *not* working within a science-unifying framework, and instead working within a balkanizing but nonetheless inclusive framework, promoting the complete opposite of unity—the "disunity of science as a working hypothesis" that was first announced by Jerry Fodor (1974) and is now so powerfully championed by Nancy Cartwright. And whichever version of the "disunity" proposal is preferred, it can be defended it on the basis of an inclusive pragmatism, which is a view that can be attractive to realists and antirealists alike.

Pragmatism is sometimes held up as deflationary of the notion of truth. Pragmatists have sometimes insisted that truth claims, as such, enjoy no special metaphysical significance: a fact is charged with whatever ontological significance it is charged with, but its truth, as such, has no more significance than that. There is nothing at all to the truth of the fact, over and above whatever happens to make it a fact. This is not to say that pragmatists have generally renounced realism. They most emphatically have not. Some pragmatists have insisted upon their claims *in service* of realism. For instance, C. S. Peirce adhered to a realist conception of truth—a conception of truth as a fixed reality, independent of human cognitions and concerns, and infused his position with a deep respect for scientific methodology: "the ideas of truth and falsehood, in their full development, appertain exclusively to the scientific method of settling opinion" (Peirce 1878). In conducting research, different investigators might use different methods to identify a quantity of interest—for example, the speed of light in a vacuum. But they will tend to arrive at the same result—because there is an independent reality at which to probe:

> So with all scientific research. Different minds may set out with the most antagonistic views, but the progress of investigation carries them by a force outside of themselves to one and the same conclusion. This activity of thought by which we are carried, not where we wish, but to a foreordained goal, is like the operation of destiny. No modification of the point of view taken, no selection of other facts for study, no natural bent of mind even, can enable a man to escape the predestinate opinion. . . .
>
> The opinion which is fated to be ultimately agreed to by all who investigate, is what we mean by the truth, and the object represented in this opinion is the real. That is the way I would explain reality. (Peirce 1992–99, 139)

I will refer to Peirce's conception as the "convergence" theory of truth.

William James, by contrast with Peirce, had a rather more forthright conception of truth that he also held to be "pragmatist":

> Any idea upon which we can ride . . .; any idea that will carry us prosperously from any one part of our experience to any other part, linking things satisfactorily, working securely, saving labor; is true for just so much, true in so far forth, true instrumentally. (1975, 34)

James's conception is thus better regarded as the "advantage" theory of truth—and it is the one that contemporary thinkers often think of as quintessentially pragmatistic.

Peirce's and James's conceptions of truth are admittedly very different. But they share a common thread—a thread that runs also through a preponderance of thinking today on the value of science: truth has a practical value; and scientific truth has an especially practical value. There is a practical benefit to truth. This aligns with the pragmatist axiom that truth makes a practical difference (and so the truth conditions to be associated with a statement are, roughly, its practical consequences).

Classical philosophical pragmatism was an American phenomenon, originating in the scholarship of the triumvirate of Peirce, James, and John Dewey, and focusing upon science as the paradigm of knowledge. Accordingly, its fundamental innovations were primarily in the area of epistemology. Fundamental too was pragmatism's commitment to Darwinism, in the form of a doctrine to the effect that the favorable designation of "knowledge" should fall only on beliefs in human history that happen to have been recruited in the service of the aim of overcoming obstacles to the satisfaction of human needs or wants. "Science," therefore must refer to those beliefs or attitudes that serve to confer upon certain organisms in the lineage of *Homo sapiens* an evolutionary advantage. Only such beliefs are eligible for the dignity of truth.

It is really quite ironic that classical pragmatism should have wished to align truth (and therefore science) with any products of natural selection. Because natural selection in no way tracks truth, but rather advantage. And sometimes that means it must promote falsehoods over truths. And for this reason certain descendants of classical pragmatism have revoked science's title to the notion of truth, altogether denying truth a place of philosophical significance, and consequently take a rather different (though not necessarily dimmer) view of science. Donald Davidson and Richard Rorty, for example, propose to dispense with the notion of truth altogether from epistemic theorizing, because it does not provide an independent standard for demarcating the wheat from the chaff, as more properly epistemic terms (such as for instance *warrant* or *justification*) might do.

Rorty especially (1991) emphatically denies that beliefs purport to portray the world "as it really is." For how could a species have acquired the ability to *represent* the universe—especially the universe as it really is, as opposed to how it is usefully described, relative to the particular needs of members of that species?

In what remains of this book, I shall be distancing my position from that of pragmatism, both classical and contemporary. I agree with contemporary pragmatists that natural selection can track useful beliefs. So I shall argue that natural selection nowise tracks truth, as such, even though it sometimes promotes truth over falsehood. But because falsehoods can be exceptionally useful, natural selection cannot help but sometimes promote these as well. And so I shall maintain that natural selection does not promote beliefs on the basis of their truth values. But *unlike* pragmatists, classical or contemporary, I shall insist upon the value of affirming truth for another reason entirely. I shall argue that truth is an epistemically *normative* notion—in other words that it marks certain propositions as deserving of belief. Indeed, the value of marking something as true is that it thereby signifies it as meritorious *independently* of how believing it might confer an advantage. In defending this view, I shall be rejecting entirely the pragmatic theory of truth and pragmatistic conceptions of science with it, and giving voice to a deeper reason for the unity of science. For it is simply *not* the case that the products of natural selection always deserve to be called Science. We should want to reserve the favorable designation of "Science" for something else. And I contend that we can turn our attention to that something else only when we view the world as unified in the way I have been describing, and therefore deserving of a science-unifying framework.

Before the book is concluded, we shall examine Bernard Williams's contention that *belief* is a normative concept: in other words that there is an intrinsic aim to belief, and that aim is truth. This is quite a problematic dictum, as it stands, insofar as a believer, when forming a given belief, may be enjoying no such aim—and may indeed be intent upon advancing aims (reputation, personal gain, simple perversity, to name just a few) that can conflict with the goal of truth. Still there is an important idea underlying Williams's dictum. I shall argue that truth itself is normative. It is the very name of truth that bears the normative property, not believers, or belief, as such. Indeed truth holds itself up as the aim of inquiry, no thanks at all to us. This is an Aristotelian idea. Still, it is a fact about humans that we subscribe to the standard that truth deserves to be believed, whether or not we aim personally at it. This normative feature of truth is the extra something that Rorty quite correctly felt was nowise a guide to action—not if the would-be believer is already committed to outing truth. And Rorty maintained—again correctly—that truth claims should play no role in evaluations of competing beliefs on a topic. And he is quite right, as that role belongs only to evidence. Moreover, instructing someone to pursue the truth (rather than the convenient, for instance) affords that person no additional help in working out what that truth might be. However, marking something as true (or, relatedly, likely to be so) provides the fundamental—perhaps the *only*—reason for utilizing information already on offer toward a certain conclusion, a reason that is emphatically *not* rooted in concerns for personal advantage.

These remarks serve to demonstrate that there is something wrong with the deflationary theory of truth. Deflationism advises, perversely, that truth as such has no claim on believers independent of their other aims. But the way we cope with statements about truth, as I shall argue, tells a different story: we are not indifferent to truth. Truth most assuredly makes claims upon us. Truth is more substantial than deflationists would have us believe. I shall argue that it is precisely because truth exercises claims upon us that we are subject to an imperative to adopt a science-unifying framework.

7. Evolution by Natural Selection Is Blind to Truth

It is truly an irony that pragmatists, classical and contemporary, view Darwinism as an ally. Richard Rorty proclaims that, like Dewey, he can discern "no breaks in the hierarchy of increasingly complex adjustments to novel stimulation—the hierarchy which has amoeba adjusting themselves to changed water temperature at the bottom, bees dancing and chess players check-mating in the middle, and people fomenting scientific, artistic, and political revolutions at the top" (Rorty 1991, 109). And this lead him to the view (which he regards as a refined form of pragmatism) that it makes no sense at all to say that humans pursue truth in the process of, instead of or simply in addition to, pursuing their advantage. This is because our vocabularies "have no more of a representational relation to an intrinsic nature of things than does the anteater's snout or the bowerbird's skill at weaving" (1998, 48). By contrast, Philip Kitcher (1993) maintains that the institutions of science were the direct result of natural selection: at the beginning, the standards and methods of inquiry were largely a matter of social factors; but eventually humans arrived at a method—familiar now as the scientific method—for resolving disputes, via the process of natural selection. This is because natural selection ensured that those who employed the method better or more perfectly prospered more than those who did not.

Both Rorty's and Kitcher's positions, however, are no more than articles of faith—Rorty's because it is quite obvious that those humans who can "track" the truth, for instance about where their prey is located, will, like any other predator, be favored by natural selection over a competitor whose grasp of prey's whereabouts is more spotty. So natural selection does track *some* features of how an animal *represents* its world. But Kitcher's position is also based on faith, because the features of a belief that natural selection tracks are in a vast number of cases not its truth. This is also easy to see. Consider the human who believes he is the most attractive male in his cohort. That male has a much better shot at out-reproducing his competitors than a comparable male minus the self-serving belief. The first male will ensure he is at the right places at the right times. But his chances of reproducing himself are not improved by the *truth* of his self-serving belief, but simply by the fact that he believes it. In fact, it might well be the comparable male who believes the truth—that indeed he is not very attractive. So, the unattractive male would be much better served by a falsehood on the subject than he is by the truth. It is no wonder, then, that so many of us believe we are better than average. And about half of us are absolutely wrong—but nowise worse off for believing a falsehood. To the contrary: the falsehood serves us considerably better than the truth in this case and many others.

Natural selection tracks advantage. When it comes to beliefs, nature will select beliefs (when it exerts pressures on beliefs at all) that serve their bearer better, all things considered, over beliefs that serve less well. And so Darwinism is no ally of the idea that human beings track truth—unless advantage and truth converge, or are artificially made to do so, such as when human beings organize their institutions accordingly (as Longino 1990 suggests). It will help to argue this point in another way as well.

In the *Republic,* Glaucon tells the story of Gyges. It is a story that deserves retelling, with modifications, now and again. This is such a time. Once upon a time a chap called Gyges found a ring with a remarkable power: when "activated," it delivers to the brain a dose of a drug that induces in the person bearing it a package of beliefs tailor-made to advancing the bearer's objectives. So for example, if Gyges is now hungry, the ring's activation will induce in him the belief (true, as it happens) that there's an injured rabbit in the bush to his right, and that if he will only take hold of that tree limb yonder, he can have a nice rabbit stew for dinner. With this ring Gyges need never go through the effort of learning to track prey, or judge a bargain, or analyze a business prospect: what he needs, by way of belief, comes to him as if he always already knew it.

Of course new beliefs require some reconciling to old ones. So the ring will induce a tendency for beliefs in its bearer to equilibriate themselves—a tendency for tensions in the corpus of beliefs to be reduced through pruning of outliers.[4] Naturally, after putting the ring to some use, Gyges will be unable to discern the origins of any given belief: thus it will be a natural outcome of multiple uses of this ring that the line between ring-induced beliefs and others will be blurred, even if in a given application the incremental change might be immediately obvious.

[4] Indeed there is a very large literature proposing alternative algorithms for belief corpus modification, upon the introduction of new propositions into a given corpus.

What will Gyges's corpus of beliefs look like after some time with the ring? Does it contain only truths? No. It is liable to contain many falsehoods. For instance, Gyges, just like the overconfident male we encountered earlier, will very likely come to believe he is uncommonly attractive. Will the contents of Gyges's corpus depend on when he started using the ring, what he believed at the time, and his special aims in life? Of course it will. But of course it would have done absent the ring. Still, however serviceable Gyges's ring-sourced convictions might otherwise be, they are not all true, and particularly not if we assume (as we ought to do) that Gyges does not use the ring solely out of idle curiosity. For there will always be a cost to forming a belief. So we should postulate that in Gyges's case the ring tires him out or is costly to his health in some way. So he cannot use the ring without counting some costs. The conclusion we can press here is that belief—as a commodity that itself draws and depends upon a certain pool of an organism's limited resources—is decisively *not* aimed at truth. Its target, instead, is advantage. And these are distinct things, pragmatist convictions notwithstanding.

It does no good to protest that by and large, and in ordinary circumstances, it will be true opinions, and not false, that Gyges' ring will foist upon him, or that the ring is probably a reliable source of true belief on many topics. For so long as there exists a single instance in which a false opinion will serve Gyges's aims better than a true one, it will be the case that the imperative to advance advantage is not the same as the imperative to pursue (or approximate or even converge upon) the truth.[5]

Gyges's ring is a technology enhancing the cognitive corpus for better service of its user's objectives. Although Gyges's ring is better than what we are presently capable of, we are all consumers of such technologies, most prominently in the form of sense organs.[6] These have been "engineered" in natural history to serve aims specific to the context of our species, whether or not we as agents embrace those aims when we employ these services today. Their purpose is not to deliver truth, but to deliver beliefs that afford an advantage. In support of this we need only note that one of the beliefs foisted upon us by our visual apparatus is that a preponderance of medium-sized objects in the world are constituted of continuous matter, which we know (because of modern physics) is false. So appealing to practicality of a belief is precisely the wrong tactic, if one wishes to advise on the truth of a belief. This point shall be for us the beginning of wisdom.

[5] Glaucon formulates his original Gyges story to secure a similar leverage against Socrates's position, and in favor of his own proposal to the effect that pursuit of self-interest and pursuit of justice are decidedly different things. The Gyges stratagem works for Glaucon the same way I should like it to work for my case in the present context.

[6] Cognitive theories of vision-for-action (for example Milner and Goodale 1995) provide strong independent evidence for this claim, though there is no space here to elaborate the point.

8. Significant Truth

Kitcher (2001, 2011) argues that science aims at "significant truth" rather than at truth unqualified. Against science skeptics, Kitcher argues that science is indeed an objective and truth-seeking enterprise; but against unmodulated science boosterism he argues that science is also interest-relative—because what we inquire into and the categories that structure scientific inquiries must be sensitive to social and political concerns. Thus what science seeks is not truth unqualified, but *significant truth.*

I can imagine someone interjecting at this point that this is enough to rescue science from the clutches of the argument of the previous section. After all, the examples of beliefs serving individual advantage so far considered, which *diverge from truth,* will all be insignificant. The overconfident male, too much impressed by his own attractiveness, does not hold a belief deserving of the dignity of the label "science"—and certainly has not reached it via the strictures of an approved method.

But this is cold comfort. So long as it is we who say what "significant" means, there is no guarantee of any sort that false beliefs that afford certain advantages will not find their way into the scientific corpus. Indeed many have already: beliefs, for instance, as to the inferiority of certain members of the human species—women, the underclasses, persons of color. Once upon a time such "truths" were considered quite significant indeed, and by scientists no less so than by ordinary mortals who might never insist upon such sentiments as "scientific." Scientists are not immune to the powers of natural selection; Science thus can be bent and abused. We have to be watchful over scientific institutions precisely because of this fact—ensuring that the right incentives are in place to help us identify the chaff sooner rather than later.

9. Aristotle's Theory of Science, Once Again

Aristotle view of science, which I introduced at the very beginning of our journey, deserves reviving. Aristotle insisted on a distinction between knowledge (science) and common sense: science is that which adheres to standards of inquiry—standards independent of the would-be believer's interests. Common sense, by contrast, answers only to its users. Aristotle also insisted on philosophy having its own standards, so that philosophical reflection on a discipline comes out as different from the practice of the discipline itself (cf. Irwin 2002).[7]

[7] This corroborates certain sentiments—those that provoked J. L. Austin, G. E. Moore, and Ludwig Wittgenstein to advance the quietist doctrine—to the effect that when one asks, "How do I know there is a hand here?" one is not seeking an ordinary answer to a question at all, but only a philosophical answer. Austin and Wittgenstein famously held that there is none to give.

Aristotle's theory of science and his conception of knowledge are one and the same. On this conception, common sense is not knowledge, because common sense enjoys no standards of inquiry. It is not a discipline; it is an activity of practical, as contrasted with scientific or disciplinary, life. Knowledge is relatively rare, at least in certain epochs, in that it is the product of science. It is common only conditionally, and common only to the extent that scientific knowledge is disseminated well and freely. If scientific inquiry had never been conducted, or if it were secreted in the temple of a holy priesthood, then there would be no knowledge spread across the earth: the planet would be covered only in abject common sense.

I want to make this perfectly clear. Aristotle's view is not a version of quietism as invented by G. E. Moore and Ludwig Wittgenstein. Quietism maintains—to the contrary of Aristotle—that common sense enjoys the highest epistemic authority. And this is why, according to the quietists, common sense bears no burden to give a special account of itself as knowledge. For it is the prototype of what knowledge is. This is the polar opposite of the view of truth and knowledge that I am proposing to revive in Aristotle's name. When the Aristotelian says that common sense has no reply to offer a challenger, this is just another way of saying that common sense does *not* meet the high standards of intellectual inquiry. Still, one need not deny that something with common-sense-like content can become scientific data or scientific observations. But this can happen only when the contents in question pass very rigorous standards of scientific observation, under the right conditions for making such observations.[8]

10. Knowledge

The moral of our adapted Gyges tale is that knowledge, on the one hand, and belief of convenience (and plain common sense too), on the other, are different things. Why go to such lengths to mark the distinction? Its truth, it would seem, is so obvious. This is not so. We need to emphasize that the kind of rationality involved in the assembly of true knowledge is considerable different from practical rationality.[9] And this idea goes counter to much that many present-day epistemologists and philosophers of science hold dear.

[8] This in not the place to say what these standards are. In any case, the matter is subject to some controversy.

[9] Recognition of this fact is one route to solving certain philosophical puzzles of recent vintage, problems having to do with strategy: puzzles of deterrence, or of coordination where there are multiple Nash equilibria, and the toxin puzzle (where prudence as such seems to get in the way). All of these problems lend themselves to better solution via Gyges's ring than through a process of ratiocination in which along the way we acquire materials that undermine a desirable target belief.

Gyges's ring allows us to enslave judgment to the passions.[10] While the ring of the present Gyges tale can foist upon its user such beliefs (many of them true) as will help meet present objectives, what will it do when the user's objective is knowledge? Can it through chemical means assemble for its user such evidence as will make the belief associated with it deserve the dignity associated with the name of knowledge? To be sure, we can grant that the ring can manufacture the *appearance* of evidence, but of course the appearance of evidence and the reality of it are different things.

When we speak of knowledge, we are talking about something of very recent provenance in natural history, supported by a multinational web of institutions, and serving a market that is markedly different from that served by the beliefs Gyges is after in our adapted story. We are indeed talking about something that aims at truth as such, without regard for its utility. To study this something in the light of natural history, we have to mark it off from the pretenders—the beliefs of convenience. Only then can we recognize that it serves a very different market, and that therefore its natural history should be quite distinct from that of convenient beliefs. Only then will we acknowledge that it must answer to the universe as it actually is.

11. The Natural History of Knowledge: Some Preliminary Gestures

Pragmatism today subdivides into a number of species, but uniting them all is the idea that science, such as it is, is "natural" in the sense that it is continuous with common sense and with what I referred to in chapter 5 as the "clinical orientation"—that science serves each human organism in the way that a bat's wing serves the bat. Pragmatists are right (once again, ironically) that science is a development in the natural order, and one deserving of a Darwinian natural history. My account differs from theirs in that it asserts the discontinuity between science and common sense, as I will now attempt to make clear.

Social life is a network of strategic interactions. It is a stage on which strategies, like everything else that is subject to transmission from one generation to the next, vie for survival and reproduction.[11] Strategies need not be individual- or organism-bound, for their transmission is not restricted (as biological transmission is) to the vertical transmission from parent to offspring. Because strategies among the likes of us can be transmitted through imitative learning,

[10] Rather more directly, I should think, than what Hume had in mind.

[11] Ever since the appearance of the influential work of John Maynard Smith (1982), this characterization of the social has been heavily articulated.

they can float freely in a culture. When related strategies coalesce, we may call the resulting package a *practice.*[12]

Now, a natural history of a practice—as a natural history—records not only the function of the practice being singled out for attention, but also the competitors against which it won the day, as well as the circumstances or evolutionary pressures that were such as to render it more likely to reproduce itself in relation to its competitors. So what were the competitors to our current knowledge practices? And what sorts of circumstances rendered their adoption more advantageous, and so the better smiled-upon by selection forces? Why is knowledge, thought of as a community's repository of carefully preserved theories and hard-won evidence for them, worth having? In other words, what market does it serve? What practices has it had to compete with or to displace, and how has it won renewed and redoubled investment? Perhaps the specifics here can never be known for certain. The only study within miles of these questions was made by Edward Craig (1990) over two decades ago and taken very little notice of. Craig asks what knowledge is for in the economy of human practices. He defends his proposed answer to this question—that knowledge practices serve the social function of marking reliable informants as knowers—by showing that, were this hypothesis true, then the practices we currently engage in, vis-à-vis exchange of propositions, would make sense. But this is not yet a natural history in the sense we have come to esteem.

Still, one thing is very clear: there is more than one practice involved in the trading of propositions—it is not all for the sake of sharing knowledge. Some practices clearly do not involve exchange of information under the banner of knowledge-sharing. These are the practices that speech act theorists have called to our attention. When I say, "I promise to repay you in a month's time," or "I take you as my husband," I am not sharing any information, even if I am saying something true or enforceable. While some practices serve individual goals, others—like the knowledge practice—serve the goals of communities. All extant practices serve a purpose or function—and the function in question is worth serving, or else the resources devoted to it would long ago have been devoted to serving some other end.

I contend that the knowledge enterprise is not continuous with common sense or the clinical orientation because we cannot give an account of common sense or the clinical attitude as aiming at truth exclusively, or as functioning to provide the believer with true belief as such. Precious resources go into the fixation of each belief. The moral of the Gyges story is that

[12] And there is a growing school of social scientists directing enormous creative energies toward founding a quantitative science of culture and anthropology in this spirit. Many of them reserve a space for cultural transmission as independent and normally also admit of vector combinations of biological and cultural forces. See, for example, Boyd and Richerson (1985) and Cavalli-Sforza and Feldman (1981). Skyrms (1996) offers a modeling strategy (called "replicator dynamics") that can be utilized to model both types of forces.

apportioning belief according to the evidence would be an evolutionary disaster when it could be apportioned instead so as to maximize advantage. In the natural world belief is one more means to sustaining the organism's biological and economic being; so that aiming at truth for its own sake is a luxury that no organism can afford. So long as there is no *independent* payoff to apportioning belief in a way that makes it likely to be true directly, and not indirectly through the truth's being a useful commodity, there can be no natural history of knowledge. But if there is to be such a payoff, it must be artificially introduced—say, through cultural institution, for it does not occur in nature in a systematic enough way—or else truth and advantage would not be divergent, which they are. Knowledge and common sense are consequently not continuous with one another: at least one speciation event must intervene. Thus the evolutionary paths to our current condition vis-à-vis those two enterprises are substantially different. On this point will hang the rest of our tale in this book. But first a gesture at the natural history of knowledge.

Common sense, as already noted, strikes a balance between self-advantage, on the one hand, and apportioning belief according to a currency exchangeable in the human marketplace for goods and services, on the other. Knowledge, by contrast, has a different provenance. Knowledge enters the stage of history when institutions are created to pay off individuals or collectives (in a currency they already accept) for argumentation supporting a proposition after a fashion that illuminates the likelihood of its truth directly. A natural history of this latter sort can be told only when collectivities and institutions *too* are allowed to play roles on the evolutionary stage.

The history I will float is (roughly) as follows: in addition to individual organisms, we also have on the evolutionary stage institutions and collectivities, such as those that disseminate ideas, publish printed matter, and compete (in various markets) for the loyalties of consumers, adherents, and students. They exert evolutionary pressures upon one another and upon individuals and are acted upon by pressures exerted by individuals. To the extent that a market can be created for true belief as such, there is a special ecological/cultural niche to be filled. And institutions can rise up to supply that need. Institutions that seek to serve that market will find themselves in competition with one another, and possibly also in competition with individuals after a share of that same market. We are calling institutions that currently occupy that niche in our present milieu by the name of Science. Science—as an institution—provides the cognitive technologies required for the manufacture of knowledge, as well as for creation of the market that the product serves. Alone, naked human minds are not up to all of that.

A working assumption I have relied upon here (as Craig did before me) is that all extant practices serve a purpose or function—and that the function in question is worth serving, or else the resources devoted to it would long ago have been devoted to serving some other end. Knowledge practices—the practices of requesting and sharing knowledge—are practices that originate from deep needs to gather information, to be sure, but those need not all be so self-serving as has been supposed in some pragmatist theories of knowledge. For knowledge is much less the sharing of common sense in service of "clinical" goals than it is the preservation of material that conforms to certain rather rigorous standards of evidence for intellectual inquiry. The natural history of such practices belongs alongside the history of institutions with a very special kind of social mission. It is thus the stuff of social science and philosophy, and not the stuff of biology alone.

12. Three Cheers for Antipragmatism: Truth Is Normative

> For though both are dear, it is our sacred duty to choose the truth in preference to our friends.
>
> —Aristotle, *Nicomachean Ethics*

Pragmatism's mistake lies in not relinquishing Bernard Williams's contention to the effect that "belief aims at truth," which rests on the assumption (false, as pragmatism itself recognized) that there can be only one aim for belief. Or at any rate, that this philosophically virtuous aim cannot conflict with more creaturely aims for advantage. The fact is that belief serves many masters. Most of these know nothing and care even less about truth. This is the very genius of truth: it is orthogonal to all such things.

When someone asserts that something is true, she is purporting that the reasons for believing it are independent of any advantage we may wish to gain thereby. She is testifying that the belief in fact lies outside of the branch of natural beliefs that serve individual advantage, via common sense. She is staking her reputation on the contention that the merits of the proposition in question do not lie in its service to us. She is instructing us that the standards to apply, when deciding whether to adopt the proposition, are the standards of science broadly construed—in Aristotle's terms, that indeed it amounts to knowledge because it can or does meet strict standards. To remark on something as true is in effect to say that it passes or can pass muster vis-à-vis standards of this sort. It is to remind people of advantage-independent standards, and to urge them to refuse to engage in advantage-based reasoning if they are susceptible to it.

The ideas of the previous paragraph have been taken up with making sociological observations, with examining relations among utterances that people are apt to make when they are involved in certain dialogical interactions—the stuff that Paul Grice (1975) might have referred to as *implicature*. This analys is is sometimes construed as examining intuitions, or looking for the meanings of words. But inquiry into implicatures is no more a matter of examining things in a linguistic speaker's head than a pH test is a test of pH paper. It is proper sociology. And in this case it reveals something about the importance of the practice of talking about truth. The language of truth in contexts where it actually erupts signifies that the speaker who employs it is invoking standards of inquiry—standards of the sort Aristotle talked about when he talked about the different sciences, the standards that mark scientific inquiry off from common sense.

The pragmatists were not wrong to associate truth with science. It is just they did not remember what Aristotle said about the foundation of science—namely that it involves application of standards. Aristotle did not shy away from norms. He thought they were part of the natural order—a view that contemporary pragmatists do not share.

To say that something is true is to gesture at an impersonal norm. This impersonal norm is as far as we humans can get from reckoning in terms of personal advantage. To us, the gesture at truth is a gesture that is meant to point us in a direction diametrically opposite to personal advantage. That is the genius of the designation "true." It signifies a rationale for accepting something that is completely independent of any aims one might have for personal gain. (It's important that the rationale be independent, and not simply *opposed* to one's aims. For of course if one knew that one's aims would be disserved by the truth, one would then be motivated to strive against it as best one could.)

To conclude, then: the pragmatists were right that belief does not aim at truth. They were right that belief is not itself normative. They drew from this the conclusion that there is nothing at all to the notion of truth. But this is incorrect. Truth, as a conception, is assuredly normative. And normativity is not nothing. To say that something is true is to mark that it passes or can pass standards of scientific inquiry, broadly construed, and for that reason deserves allegiance. Truth deserves to be believed. And saying so is—just as it seems—something of a tautology.

This shows that there is something wrong with the deflationary theory of truth. Truth has a claim on believers, whether or not they aim at believing truth. That is the genius of truth. Truth, as such, bears the normative property. *It* is the thing deserving—not individual propositions, as such. Truth is thus considerably more substantial than deflationists would have us believe. The assertion of truth, in any context of utterance, brings into the context the suggestion that norms of inquiry must govern the proceedings from that moment forward. It reminds participants that there are imperatives that deserve the allegiance of those engaged in the enterprise—whether or not they are internally motivated to comply with those imperatives. "Truth" signals a context of imperatives above and beyond loyalties to personal goals. That is the dignity of truth.

The dignity of Truth is the reason that Science aims at it. That is Science's own dignity. And because Science aims at Truth, it must utilize a unifying framework, since the world—as this book has argued—is itself unified.

REFERENCES

Abraham, R. and Shaw, C. D. 1992. *Dynamics: The Geometry of Behavior*. Aerial Press, Santa Cruz, CA, 2nd edition.

Albert, D. Z. 1992. *Quantum Mechanics and Experience*. Harvard University Press, Cambridge.

Alexander, S. 1920. *Space, Time, and Deity: The Gifford lectures at Glasgow, 1916–1918*. Macmillan, London. 2 volumes.

Anderson, P. W. 1997. *Concepts in Solids: Lectures on the Theory of Solids*. World Scientific, Singapore. Originally published in 1963.

Auyang, S. L. 1998. *Foundations of Complex-System Theories: In Economics, Evolutionary Biology, and Statistical Physics*. Cambridge University Press, New York.

Barabási, A.-L. 2003. *Linked: How Everything is Connected to Everything Else and What it Means for Business, Science, and Everyday Life*. Penguin, New York.

Bardeen, J., Cooper, L. N., and Schrieffer, J. R. 1957a. Microscopic theory of superconductivity. *Physical Review* 106:162–64.

Bardeen, J., Cooper, L. N., and Schrieffer, J. R. 1957b. Theory of superconductivity. *Physical Review* 108(5): 1175–204.

Barnes, B. 1995. *The Elements of Social Theory*. Princeton University Press, Princeton, NJ.

Batterman, R. W. 1995. Theories between theories: Asymptotic limiting intertheoretic relations. *Synthese* 103(2): 171–201.

Batterman, R. W. 1998. Why equilibrium statistical mechanics works: Universality and the renormalization group. *Philosophy of Science* 65(2): 183–208.

Batterman, R. W. 2000. Multiple realizability and universality. *British Journal for the Philosophy of Science* 51(1): 115–45.

Batterman, R. W. 2002. *The Devil in the Details: Asymptotic Reasoning in Explanation, Reduction, and Emergence*. Oxford Studies in Philosophy of Science. Oxford University Press, New York.

Batterman, R. W. 2008. Intertheory relations in physics. *Stanford Encyclopedia of Philosophy*. Zalta, E. N., editor, Fall 2008 edition.

Bealer, G. mss. Analysis and definition.

Bechtel, W. and Richardson, R. C. 1992. Emergent phenomena and complex systems. In Beckermann, A., editor, *Emergence or Reduction*, pages 257–88. de Gruyter, New York.

Bechtel, W. and Richardson, R. C. 1993. *Rediscovering Complexity*. Princeton University Press, Princeton, NJ.

Bedeau, M. 1997. Weak emergence. In Tomberlin, J., editor, *Philosophical Perspectives 11: Mind, Causation, and the World*, pages 773–840. Blackwell Publishers, Oxford.

Bell, J. S. 1964. On the Einstein Podolsky Rosen paradox. *Physics* 1:195–200.

Bickle, J. 2011. Multiple realizability. *Stanford Encyclopedia of Philosophy*. Zalta, E. N., editor, Winter 2011 edition.

Binney, J. J., Dowrick, N. J., Fisher, A. J., and Newman, M. 1992. *The Theory of Critical Phenomena: An Introduction to the Renormalization Group*. Oxford University Press, Oxford.
Bonner, J. T. 2011. Why Size Matters. Princeton University Press, New Jersey.
Boyd, R. and Richerson, P. J. 1985. *Culture and the Evolutionary Process*. University of Chicago Press, Chicago.
Broad, C. D. 1925. *The Mind and its Place in Nature*. Harcourt, Brace, New York.
Broer, H. W. and Takens, T., editors 1992. *Geometry and Analysis in Non-linear Dynamics*. Wiley, New York.
Bunge, M. A. 1979. *A World of Systems*. Reidel, Dordrecht, Holland.
Bunge, M. A. 2004. How does it work? *Philosophy of the Social Sciences* 34:182–210.
Carnap, R. 1928. *The Logical Construction of the World*. University of California Press, Berkeley. Translated by George, R. (1967).
Cartwright, N. 1989. *Nature's Capacities and Their Measurement*. Oxford University Press, Oxford.
Cartwright, N. 2006a. From metaphysics to method: Comments on manipulability and the causal Markov condition. *British Journal for the Philosophy of Science* 51(1): 197–218.
Cartwright, N. 2006b. Where is the theory in our "theories" of causality? *Journal of Philosophy* 103(2): 55–66.
Cat, J. 2009. The unity of science. *Stanford Encyclopedia of Philosophy*. Zalta, E. N., editor, Spring 2009 edition.
Cavalli-Sforza, L. L. and Feldman, M. W. 1981. *Cultural Transmission and Evolution: A Quantitative Approach*. Monographs in Population Biology, volume 16. Princeton University Press, Princeton, NJ.
Chester, M. 2003. *Primer of Quantum Mechanics*. Dover Publications, Mineola, NY.
Collingwood, R. G. 1927. *Speculum Mentis, or the Map of Knowledge*. Clarendon Press, Oxford.
Cooper, L. N. 1956. Bound electron pairs in a degenerate Fermi gas. *Physical Review* 104(4): 1189–90.
Craig, E. 1990. *Knowledge and the State of Nature: An Essay in Conceptual Analysis*. Oxford University Press, New York.
Craver, C. F. 2001. Role functions, mechanisms, and hierarchy. *Philosophy of Science* 68(1): 53–74.
Cushing, J. T. and McMullin, E., editors 1989. *Philosophical Consequences of Quantum Theory*. Notre Dame University Press, Notre Dame, IN.
Davies, P. C. W. 2006. The physics of downward causation. In Clayton, P. and Davies, P. C. W., editors, *The Re-emergence of Emergence*, pages 35–52. Oxford University Press, New York.
De Morgan, A. 1872. *A Budget of Paradoxes*. Longmans, Green, London.
Dingle, H. 1952. The nature of scientific philosophy. In *The Scientific Adventure*, page 210. Pitman and Sons, London.
Dirac, P. A. M. 1929. Quantum mechanics of many-electron systems. *Proceedings of the Royal Society of London, Series A* 123(792): 714–33.
Dyson, F. J. and Lenard, A. 1967. Stability of matter pt. I. *Journal of Mathematical Physics* 8:423–34.
Dyson, F. J. and Lenard, A. 1968. Stability of matter pt. II. *Journal of Mathematical Physics* 9 (5): 698–711.
Elster, J. 1989. *Nuts and Bolts for the Social Sciences*. Cambridge University Press, New York.
Epstein, B. 2011. Agent-based models and the fallacies of individualism. In Humphreys, P. and Imbert, C., editors, *Models, Simulations, and Representations*, pages 115–44. Routledge, New York.
Fine, K. 1991. A study of ontology. *Noûs* 25:263–94.
Fine, K. 1994. Ontological dependence. *Proceedings of the Aristotelian Society* 95:269–90.
Fine, K. 2010. Some puzzles of ground. *Notre Dame Journal of Formal Logic* 51(1): 97–118.
Fine, K. forthcoming. The pure logic of ground. *Review of Logic*.
Fodor, J. A. 1974. Special sciences. *Synthese* 28(2): 97–115.
Fodor, J. A. 1997. Special sciences: Still autonomous after all these years. *Philosophical Perspectives* 11:149–63.
Fraiberg, S. and Adelson, E. 1977. Self-representation in language and play. In Fraiberg, S., editor, *Insights from the Blind*, pages 248–70. Souvenir, London.

Glymour, C., Scheines, R., and Spirtes, P. 2001. *Causation, Prediction, and Search*. MIT Press, Cambridge, 2nd edition.

Glymour, C., Spirtes, P., and Scheines, R. 1992. *Causation, Prediction, and Search*. Springer-Verlag, New York.

Gould, S. J. 1999. *Rocks of Ages*. Library of Contemporary Thought. Ballantine Publishing Group, New York.

Greetis, E. A. mss. Why "autonomy solutions" cannot avoid Kim's downward causation argument.

Grelling, K. 1988. A logical theory of dependence. In Smith, B., editor, *Foundations of Gestalt Theory*, 218–27. Philosophica Verlag, Munich.

Grice, H. P. 1975. Logic and conversation. In Cole, P. and Morgan, J., editors, *Syntax and Semantics 3: Speech Acts*, pages 41–58. Academic Press, New York.

Grimes, T. 1988. The myth of supervenience. *Pacific Philosophical Quarterly* 69:152–60.

Happé, F. and Ronald, A. 2008. The "fractionable autism triad": A review of evidence from behavioral, genetic, cognitive and neural research. *Neuropsychology Review* 18(4): 287–304.

Happé, F., Ronald, A., and Plomin, R. 2006. Time to give up on a single explanation for autism. *Nature Neuroscience* 9(10):1218–20.

Hardcastle, V. G. 1995. A critique of information processing theories of consciousness. *Minds and Machines* 5(1): 89–107.

Hawking, S. W. 1988. *A Brief History of Time: From the Big Bang to Black Holes*. Bantam Books, New York.

Hedström, P. and Swedberg, R. 1998. *Social Mechanisms: An Analytical Approach to Social Theory*. Cambridge University Press, New York.

Henry, O. 1982. *The Gift of the Magi*. Neugebauer Press. Originally published in 1906.

Hitchcock, C. 2001a. The intransitivity of causation revealed in equations and graphs. *Journal of Philosophy* 98(6): 273–99.

Hitchcock, C. 2001b. A tale of two effects. *Philosophical Review* 110(3): 361–96.

Hitchcock, C. 2007. Probabilistic causation. *Stanford Encyclopedia of Philosophy*. Zalta, E. N., editor, Fall 2007 edition.

Hobson, R. P. 1993a. *Autism and the Development of Mind*. Lawrence Erlbaum Associates, Hillsdale, NJ.

Hobson, R. P. 1993b. Through feeling and sight to self and symbol. In Neisser, U., editor, *The Perceived Self*, pages 254–79. Cambridge University Press, New York.

Hobson, R. P. and Bishop, M. 2003. The pathogenesis of autism: Insights from congenital blindness. In Frith, U. and Hill, E., editors, *Autism: Mind and Brain*, pages 109–126. Oxford University Press, New York.

Hume, D. 2009. *An Enquiry Concerning Human Understanding*. In Ariew, R. and Watkins, E., editors, *Modern Philosophy*, pages 533–600. Hackett, Indianapolis, IN. Originally published in 1739.

Humphreys, P. 1997a. Emergence, not supervenience. *Philosophy of Science* 64(4): 337–45.

Humphreys, P. 1997b. How properties emerge. *Philosophy of Science* 64(1): 1–17.

Hüttemann, A. 2004. *What's Wrong with Microphysicalism?* Routledge, New York.

Irwin, T. 2002. Aristotle. *Routledge Encyclopedia of Philosophy*. Routledge, New York.

James, W. 1975. *Pragmatism: A New Name for Some Old Ways of Thinking*. Harvard University Press, Cambridge. Originally published in 1907.

Kadanoff, L. P. 1966. Scaling laws for Ising models near Tc. *Physics* 2(263).

Kanner, L. 1943. Autistic disturbances of affectice contact. *Nervous Child* 2:217–50.

Khinchin, A. I. 1949. *Mathematical Foundations of Statistical Mechanics*. Dover Publications, New York. Translated by Gamow, G.

Kim, J. 1993. Supervenience as a philosophical concept. In *Supervenience and Mind: Selected Philosophical Essays*, pages 131–60. Cambridge University Press, Cambridge. Cambridge Studies in Philosophy.

Kim, J. 1999. Making sense of emergence. *Philosophical Studies* 95(1): 3–36.

Kitcher, P. 1993. *The Advancement of Science: Science without Legend, Objectivity without Illusions*. Oxford University Press, New York.

Kitcher, P. 2001. *Science, Truth, and Democracy*. Oxford University Press, Oxford.
Kitcher, P. 2011. *Science in a Democratic Society*. Prometheus Books, Amherst, NY.
Koslicki, K. 2008. *The Structure of Objects*. Oxford University Press, New York.
Lanczos, C. 1949. *The Variational Principles of Mechanics*. University of Toronto Press, Toronto.
Langlois, R. 1983. Systems theory, knowledge, and the social sciences. In Machlup, F., editor, *The Study of Information: Interdisciplinary Messages*, pages 581–600. John Wiley, New York.
Lenard, A. and Dyson, F. J. 1968. Stability of matter pt. II. *Journal of Mathematical Physics* 9:698–711.
Levins, R. 1970. Complexity. In Waddington, C. H., editor, *Organization Stability and Process: Toward a Theoretical Biology*, vol. 3, pages 67–86. University of Edinburgh Press, Edinburgh.
Lewis, D. 1973. Causation. *Journal of Philosophy* 70:556–67.
Lewis, D. 1979. Counterfactual dependence and time's arrow. *Noûs* 13:455–76.
Little, D. 1998. *Microfoundations, Methos, and Causation: On the Philosophy of the Social Sciences*. Transaction Publishers, New Brunswick, NJ.
Longino, H. E. 1990. *Science as Social Knowledge: Values and Objectivity in Scientific Inquiry*. Princeton University Press, Princeton, NJ.
Machamer, P., Darden, L., and Craver, C. F. 2000. Thinking about mechanisms. *Philosophy of Science* 67:1–25.
Mackie, J. L. 1980. *The Cement of the Universe: A Study of Causation*. Clarendon Press, Oxford.
Mainzer, K. 1994. *Thinking in Complexity: The Complex Dynamics of Matter, Mind, and Mankind*. Springer-Verlag, Berlin.
Maynard Smith, J. 1982. *Evolution and the Theory of Games*. Cambridge University Press, New York.
McGarity, T. O. and Wagner, W. E. 2008. *Bending Science: How Special Interests Corrupt Public Health Research*. Harvard University Press, Cambridge.
McGivern, P. 2008. Reductive levels and multi-scale structure. *Synthese* 165(1): 53–74.
Mellor, D. H. 1988. On raising the chances of effects. In Fetzer, J. H., editor, *Probability and Causality*, pages 229–39. Kluwer, Dordrecht, Netherlands.
Mermin, N. D. 1981. Quantum mysteries for anyone. *Journal of Philosophy* 78(7): 397–408.
Mill, J. S. 1874. *A System of Logic, Ratiocinative and Inductive: Being a Connected View of the Principles of Evidence and the Methods of Scientific Investigation*. Harper and Brothers, New York, 8th edition.
Milner, D. and Goodale, M., 1995. *The Visual Brain in Action*. Oxford University Press, New York.
Mitchell, M. 2009. *Complexity: A Guided Tour*. Oxford University Press, New York.
Mitchell, S. D. 2009. *Unsimple Truths*. University of Chicago Press, Chicago.
Nagy, M., Akos, Z., Biro, D., and Vicsek, T. 2010. Hierarchical group dynamics in pigeon flocks. *Nature* 464(7290): 890–93.
Newton, I. 1718. *Opticks: Or, A Treatise of the Reflections, Refractions, Inflections and Colours of Light*. Second edition, with additions. Royal Society, London
Newton, I. 1952. *Philosophiae Naturalis Principia Mathematica*. William Benton, Chicago. Translated by Adler, M. J., edited by Motte, A. Originally published in 1687.
Norman, D. A. and Shallice, T. 1986. Attention to action: Willed and automatic control of behavior. In Davidson, R., Schwartz, G., and Shapiro, D., editors, *Consciouness and Self-Regulation: Advances in Research and Theory*. Plenum Press, New York.
O'Connor, T. 2000. Causality, mind, and free will. *Philosophical Perspectives* 14:105–17.
Oliver, J. E. 2006. *Fat Politics: The Real Story behind America's Obesity Epidemic*. Oxford University Press, New York.
Pearl, J. 1988. *Probabilistic Reasoning in Intelligent Systems: Networks of Plausible Inference*. Morgan Kaufman Publishers, San Mateo, CA.
Peirce, C. S. 1878. How to make our ideas clear. *Popular Science Monthly* 12, 286–302.
Peirce, C. S. 1992–99. *The Essential Peirce: Selected Philosophical Writings*. Indiana University Press, Bloomington. 2 volumes. Edited by the Peirce Edition Project.
Pius, X. 1950. Humani generis. Encyclical letter.
Popper, K. 1959. *The Logic of Scientific Discovery (Translation of Logik der Forschung)*. Hutchinson, London. Originally published in 1935.
Reichenbach, H. 1956. *The Direction of Time*. University of California Press, Los Angeles.

Reichenbach, H. 1958. *The Philosophy of Spacetime*. Dover, New York.
Reynolds, C. 1987. Flocks, herds, and schools. *Computer Graphics* 21(4): 25–34.
Rorty, R. 1991. *Objectivity, Relativism, and Truth: Philosophical Papers*, volume 1. Cambridge University Press, Cambridge.
Rorty, R. 1998. *Truth and Progress: Philosophical Papers*, volume 3. Cambridge University Press, Cambridge.
Rosen, G. 2010. Metaphysical dependence: Grounding and reduction. In Hale, B. and Hoffman, A., editors, *Modality: Metaphysics, Logic and Epistemology*, pages 109–36. Oxford University Press, New York.
Roth, P. and Rychman, T. 1995. Chaos, Clio, and scientific illusions of understanding. *History and Theory* 34(1): 30–44.
Rueger, A. 2000. Physical emergence, diachronic and synchronic. *Synthese* 124(3): 297–322.
Russell, B. 1913. On the notion of cause. *Proceedings of the Aristotelian Society* 13:1–26.
Sawyer, K. 2001. Emergence in sociology. *American Journal of Sociology* 107:551–85.
Sawyer, K. 2002. Nonreductive individualism, part 1. *Philosophy of Social Sciences* 32:537–59.
Schaffer, J. 2000. Trumping preemption. *Journal of Philosophy* 97:165–81.
Schaffer, J. 2003. Is there a fundamental level? *Noûs* 37(3): 498–517.
Schaffer, J. 2009. On what grounds what. In Chalmers, D. J., Manley, D., and Wasserman, R., editors, *Metaphysics: New Essays on the Foundations of Ontology*, pages 347–83. Oxford University Press, New York.
Schaffer, J. mss. Grounding as the primitive concept of metaphysical structure.
Scott, A. 1995. *Stairway to the Mind: The Controversial New Science of Consciousness*. Copernicus, New York.
Sheldrake, R., McKenna, T., Abraham, R., and Houston, J. 2001. *Chaos, Creativity and Cosmic Consciousness*. Park Street Press, Rochester, VT.
Simon, H. A. 1962. The architecture of complexity. *Proceedings of the American Philosophical Society* 106(6): 467–82.
Simon, H. A. 1973. The organization of complex systems. In Pattee, H., editor, *Hierarchy Theory: The Challenge of Complex Systems*, pages 1–27. Braziller, New York.
Simons, P. 1988. Gestalt and functional dependence. In Smith, B., editor, *Foundations of Gestalt Theory*, pages 158–90. Philosophia Verlag, Munich.
Sinai, Y. G. 1982. *Theory of Phase Transitions: Rigorous Results*. Pergamon Press, Oxford.
Sinai, Y. G. 1992. *Probability Theory: An Introductory Course*. Springer-Verlag, Berlin. Translated by Haughton, D.
Skyrms, B. 1996. *Evolution of the Social Contract*. Cambridge University Press, Cambridge.
Smith, B. 2008, January 6. Barry Smith on Wittgenstein's conception of philosophy. From Warburton, N. and Edmonds, D., hosts, *Philosophy Bites*. Institute of Philosophy.
Smith, B. 1988. Practices of art. In Nyiri, J. C. and Smith, B., editors, *Practical Knowledge*, pages 172–209. Croom Helm, New York.
Sokolowski, R. 1992. Making distinctions. *Review of Metaphysics* 32:639–76.
Strogatz, S. 2003. *Sync: How Order Emerges from Chaos in the Universe, Nature, and Daily Life*. Hyperion Books, New York.
Teller, P. 1992. A contemporary look at emergence. In Beckerman, A., Flohr, H., and Kim, J., editors, *Emergence or Reduction*, pages 139–53. de Gruyter, New York.
Thalos, M. 1999a. Degrees of freedom: An essay on competitions between micro and macro in mechanics. *Philosophy and Phenomenological Research* 59:1–39.
Thalos, M. 1999b. In favor of being only Humean. *Philosophical Studies* 93:265–98.
Thalos, M. 2002a. Explanation is a genus: An essay on the varieties of scientific explanation. *Synthese* 130(3): 317–54.
Thalos, M. 2002b. The reduction of causal processes. *Synthese* 131(1): 99–128.
Thalos, M. 2003. The reduction of causation. In Thalos, M. and Kyburg, H. E. Jr., editors, *Probability is the Very Guide of Life: The Philosophical Uses of Chance*, pages 295–330. Open Court, Chicago.

Thalos, M. 2006. Nonreductive physics. *Synthese* 149(1): 133–78.
Thalos, M. 2007. Sources of behavior: Towards a control account of agency. In Ross, D. and Spurrett, D., editors, *Distributed Cognition and the Will*, pages 123–67. MIT Press, Cambridge.
Thalos, M. 2008. Two conceptions of collectivity. *Journal of the Philosophy of History* 2:83–104.
Thalos, M. 2009. Systems. *Monist* 92(3): 452–78.
Thewlis, J., Glass, R., Hughes, D., and Meetham, A. 1961. *Encyclopaedic Dictionary of Physics: General, Nuclear, Solid State, Molecular, Chemical, Metal and Vacuum Physics, Astronomy, Geophysics, Biophysics and Related Subjects*. Pergamon Press, Oxford.
Tönnies, F. 1957. *Community and Society: Gemeinschaft und Gesellschaft*. Michigan State University Press, East Lansing. translated and edited by Charles P. Loomis. Originally published in 1887.
van Brakel, J. 2000. *Philosophy of Chemistry*. Leuven University Press, Leuven, Belgium.
van Inwagen, P. 1990. *Material Beings*. Cornell University Press, Ithaca, NY.
Wiener, N. 1948. *Cybernetics: or Control and Communication in the Animal and the Machine*. MIT Press, Cambridge.
Williams, B. 1973. Deciding to believe. In *Problems of the Self*, pages 136–51. Cambridge University Press, Cambridge.
Wilson, J. 2005. Supervenience-based formulations of physicalism. *Noûs* 39: 426–59.
Wilson, K. G. 1975. The renormalization group: Critical phenomena and the Kondo problem. *Review of Modern Physics* 47(4): 773–840.
Wilson, K. G. and Kogut, J. 1974. The renormalization group and the e expansion. *Physics Reports* 12(2): 75–199.
Wimsatt, W. C. 1976. Forms of aggregativity. In Donagan, B., Perovich, A., and Wedin, M., editors, *Human Nature and Natural Knowledge*, pages 259–91. Reidel, Dordrecht, Netherlands.
Wimsatt, W. C. 2006. Aggregate, composed, and evolved systems: Reductionistic heuristics as means to more holistic theories. *Biology and Philosophy* 21(5): 667–702.
Wimsatt, W. C. 2007. *Re-engineering Philosophy for Limited Beings: Piecewise Approximations to Reality*. Harvard University Press, Cambridge.

INDEX

www.ingramcontent.com/pod-product-compliance
Ingram Content Group UK Ltd.
Pitfield, Milton Keynes, MK11 3LW, UK
UKHW021135160926
475595UK00020B/70